CANOPY CITIES

This book provides a comprehensive overview of the essential role of trees and forests in cities and examines the creative approaches cities around the world are taking to protect trees and expand their urban forests.

Moving beyond the view that trees are luxuries and therefore non-essential to the life of a city, the book examines urban tree policies and approaches that foster tree protection, including tree codes and bylaws, and calls for greater community engagement to preserve this important facet of urban life. Through an international range of examples and case studies, featuring cities in the United States, Canada, Singapore, the Netherlands, Australia, France, New Zealand, Mexico, Sierra Leone, and the United Kingdom. The book offers best practice examples where trees have been further integrated into the fabric of urban planning and design, including forested towers, interior rainforests, tiny urban forests, and metropolitan forests.

Written by a leading authority in the field, this is a fascinating read for researchers, students, and practitioners in urban planning, landscape architecture, and environmental policy and planning.

Timothy Beatley is the founder and executive director of Biophilic Cities, which coordinates a global network of partner cities working collectively to conserve and celebrate nature in urban spaces. He is a professor of sustainable communities in the Department of Urban and Environmental Planning, School of Architecture at the University of Virginia.

CANOPY CITIES

Protecting and Expanding Urban Forests

Timothy Beatley

NEW YORK AND LONDON

Designed cover image: Timothy Beatley

First published 2024
by Routledge
605 Third Avenue, New York, NY 10158

and by Routledge
4 Park Square, Milton Park, Abingdon, Oxon, OX14 4RN

Routledge is an imprint of the Taylor & Francis Group, an informa business

© 2024 Timothy Beatley

Library of Congress Cataloguing-in-Publication Data
Names: Beatley, Timothy, 1957- author.
Title: Canopy cities : protecting and expanding urban forests / Timothy Beatley.
Description: Milton Park, Abingdon, Oxon ; New York, NY : Routledge, 2024. | Includes bibliographical references and index. |
Identifiers: LCCN 2023035869 (print) | LCCN 2023035870 (ebook) | ISBN 9781032455129 (hardback) | ISBN 9781032455112 (paperback) | ISBN 9781003377344 (ebook)
Subjects: LCSH: Urban forestry. | Urban forestry--Case studies. | Trees in cities. | City planning.
Classification: LCC SB436 .B43 2024 (print) | LCC SB436 (ebook) | DDC 635.9/77--dc23/eng/20231003
LC record available at https://lccn.loc.gov/2023035869
LC ebook record available at https://lccn.loc.gov/2023035870

ISBN: 978-1-032-45512-9 (hbk)
ISBN: 978-1-032-45511-2 (pbk)
ISBN: 978-1-003-37734-4 (ebk)

DOI: 10.4324/9781003377344

Typeset in Sabon
by MPS Limited, Dehradun

CONTENTS

FIGURES

TABLES

ACKNOWLEDGMENTS

There are many individuals who deserve acknowledgment in the research and writing of this book. Thanks to the many individuals who took time for interviews by phone or zoom, and special thanks to the partner cities in our Biophilic Cities Network, who have provided many of the specific examples discussed here and who continue to inspire our work around biophilic urbanism. These include, especially: Austin, TX; San Francisco, CA; Los Angeles, CA; Portland, OR; Toronto, ON; Raleigh, NC; Richmond, VA; Reston, VA; Washington, DC; Singapore; Wellington, NZ; and Pittsburgh, PA, among many others.

But a number of other cities, not yet in our network, have provided inspiration and positive (and sometimes negative) examples. This list includes cities like Seattle, Rotterdam, Sante Fe, Freetown, and many others. What is impressive is just how many individuals and organizations in cities around the country and the world, as well as staff in parks and planning agencies, are working on tree and forest canopy protection. There is a large army of citizens and public officials working on behalf of urban trees, often faced with limited resources and many opposing forces, and that is encouraging in this era of climate change, deforestation, and global biodiversity loss. I have had the chance through this book to talk directly with many of these individuals and organizations.

And as the reader will learn, there are many specific examples and stories from Charlottesville, Virginia, a relatively small city, home to the University of Virginia. This small city is, I believe, a microcosm of many other cities and towns facing similar questions about how to manage its trees and forest canopy. A sophisticated and highly educated community, with a liberal and progressive politics that values nature and the environment, still has had a remarkably difficult time protecting and managing its trees. As I have gotten further into becoming a tree advocate in my home city, I have been startled by just how little tangible progress has been made here and how vulnerable (and diminished every day) our urban forests become. It must be worse, and more difficult, in cities without a world-class university and without the resources and supposed enlightened view of the world.

There are many specific individuals who deserve recognition for their guidance and help, too many to mention here. I thank all of the colleagues who have offered ideas and guidance over the course of a now more than thirty-year academic career.

Lastly, abundant gratitude goes to my wife Anneke, and children Carolena and Jadie, who have accompanied me on many of these forest adventures and have been the constant source of support and inspiration along the way.

PREFACE: GROWING UP IN THE URBAN FOREST

Timothy Beatley

This book is a natural culmination of multiple decades of work focused on the role of nature in cities. But my interest in the topic goes back much further. The house I grew up in sits in a forest, and thus I was immersed in the woods from an early age (See Figure 0.1). My parents were tree stewards and my father especially an advocate for saving every tree he could. We had no air conditioning in this house in the woods—we had a natural air conditioning system my dad would tell often tell me—in the form of the trees that would enwrap our home. The many screened windows allowed breezes and myriad sounds of the forests into the house. Each night I was I was lulled to sleep by crickets, katydids and tree frogs.

There was then, of course, no mention of, or awareness about, climate change, only the need for practical adaptation to the hot and humid summer months of the eastern US. I loved those summers where I would camp in the backyard. The evening sounds of the forest and pond at the bottom of the hill were mesmerizing. With the summer foliage it felt even more of a protective, magical place, and less traffic noise from the nearby highway.

The house moreover was oriented to the west, looking over and down onto a larger forested area. These woods, the setting for many days of curious exploration. There were lots of critters to see and look for here—birds, frogs and especially many box turtles—with whom I shared this living landscape with. I had at least three tree houses that I can remember, at least one built by my dad when I was quite young. They were places to escape to, to think and ponder in, to watch and listen from.

FIGURE 0.1 The Author's Childhood Home in a Forest in a City. Photo Credit: Tim Beatley.

There were other places beyond my home woods to explore and other forested areas to be found on the west end of Alexandria. The great walkability of this city meant that it was possible to get to many other groves, including the riparian forests of Holmes Run, an urban stream only a ten minute's walk away from my home.

Over the years I have had the pleasure and privilege of seeing firsthand the trees and forests of many parts of the world. I have grown to appreciate and marvel over the diversity of trees and yet the many different expressions of beauty these trees convey. Traveling has taken me to the Atlantic Forest of Brazil, the towering Jarrah-Karri forests of South West Australia, the more urban forests of Singapore, and the remnants of ancient oak woodlands of the UK and other parts of Europe.

Today for many of us our lives are made more enjoyable and tolerable because of the trees around us. As we walk our dogs, even in the morning, we find ourselves hopping from one patch of tree shade to another. The experience of walking in the sun is in many cities not just unpleasant but can be dangerous. In an era of climate change, there are few answers as unequivocal as protecting the trees we have and doing what we can to plant trees and forests for the future. It is what we must do, what we are ethically obligated to do for the many who will follow behind us.

It is hard to overstate the importance of these forests both for economic and climate protection resources but also as sources of beauty and inspiration in our lives. But especially today we recognize that trees and forests are threatened everywhere, and we cannot simply assume that they will continue as a presence in our lives. Active work to conserve, manage

FIGURE 0.2   The Flower of the Southern Magnolia Tree. Photo Credit: Tim Beatley.

and grow forests in cities is required. There are many obstacles and challenges, but there are also many wonderful ideas and creative approaches, and much effort and inspired dedication on the part of public officials, planners, citizens and others to protect what we have and to restore and renew of our urban forests for the future. This book tells many of these stories, hopefully in interesting and compelling ways.

These are daunting and discouraging times, as I watch the world literally burning up in the summer of 2023 while writing this book. It is the hottest summer on record and going to get worse. At a time when we need them most, many of our older and most precious trees, giant sequoias and bristlecone pines, are threatened by climate change. Countering this pessimistic view are the many impressive things going on in cities to raise awareness about trees and to actively protect and grow the urban canopy. There is consensus as there has never been before about the essential importance of trees, especially in cities. And from Beijing to Madrid there are tree planting and urban forest projects are underway at an unprecedented scale. There is reason, then, to be hopeful about the future, and trees and forests are a major reason (for me at least).

The chapters to follow are also meant to stand as a hopeful antidote, as well as a menu of ideas, for other cities facing similar challenges, and aspiring to a more forested future (Figure 0.2).

1

WHY ARE TREES SO IMPORTANT IN CITIES?

Urban Life Under the Shelter of a Living Canopy

We now know how many species of trees exist on the planet, roughly, thanks to a study published in 2022. The number is a remarkable 73,000, with the authors estimating that some 9000 of these are species actually yet to be discovered.[1] The remarkable diversity of tree species is a large part of what makes our planet beautiful and ecologically productive and this diversity is on display every day in and near to cities: the soaring redwoods in the San Francisco Bay area, acacia trees in Johannesburg, South Africa, the boundary oaks of London, banksia trees in Perth, Australia, and the white oaks of Washington, DC, and many other eastern US cities. They define the places where we live, and add immeasurably to urban environments where the majority of us now reside. They add immensely to our quality of life, to the everyday delight of urban life, and as it turns out are an essential part of the response to the existential threat of climate change.

In my own case, I grew up in a house in a city surrounded by trees and a forest. The house, perched on a hill, was nestled in the woods. We had no air conditioning, but the home was shaded and cooled by these trees; they also supplied abundant interactions with the many nonhuman lives that occupied these trees, especially birds. The forest was an endlessly interesting place to explore. The trees themselves formed a vertical playground for me growing up, with several tree houses where I spent much time. The trees changed with the seasons, helped me mark time, and served as a continuous source of wonder: bending and swaying (but rarely falling) in wind and storms, the leafless limbs in winter serving to frame our views of the sky. In every sense trees define our place. Without trees it is hard, really, to know where we are.

DOI: 10.4324/9781003377344-1

We have always needed trees. They provide shelter and safety, food and sustenance, materials for buildings and for burning to keep us warm, and habitat for the many other forms of life we live in close contact with. But today in the era of climate change trees are even more essential. The world is getting hotter, and cities even more so. Trees through their shade and evapotranspiration provide essential cooling benefits.

And they add an important element of beauty to our neighborhoods and our lives. Our memories are enshrined in these living monuments. They stand guard. They witness. They represent the generations of humans who preceded us and stand in for the many, many generations (hopefully) yet to live and to be sheltered and inspired by trees and forests.

Ecological Services of Trees and Forests

Trees and urban forests of course provide many ecological functions and benefits. They collect and retain stormwater, control urban flooding, they (again) cool cities, they help to filter and moderate air pollutants, and civilize cities. The scientific evidence of the health benefits of trees and urban forests continues to increase. Studies demonstrate a remarkable range of positive outcomes and impacts associated with the presence of trees in cities.

Trees and forests also play an important role in sequestering carbon, an important factor in climate change. In cities we have tended to underestimate the amount of carbon capture that can occur, especially in older trees. Planting new young trees will be important but increasingly the evidence suggests that older trees ("large diameter trees" as they are sometimes referred to) "store disproportionately massive amounts of carbon."[2]

Rising urban heat is an especially serious challenge for cities. Cities around the US and the world are already experiencing extreme heat, with serious public health impacts, and these effects will continue to get worse with climate change. Trees and forests are an unusually potent response to urban heat and while there are other strategies in the urban arsenal, such as high-albedo surfaces, few are as effective as trees. A 2017 study commissioned by the Texas Trees Foundation, found that planting trees in Dallas could result in as much as a 15 degrees (Fahrenheit) drop during the hot summer months.[3] The report concludes that these cooling effects would result from planting as few as 250,000 trees in the right places and that the trees will be some three and a half times more effective at cooling than use of reflective pavement or other cool materials strategies (though these are things we should be doing in cities as well).

Urban heat is a serious public health threat and so in many cities protecting existing trees and planting new ones will actually save lives. A recent study of 93 cities in Europe published in the Lancet estimated that if these cities were able to increase their tree canopy to 30% it would reduce premature deaths from summer heat by about 40%.[4]

Trees and urban forests are absolutely essential elements of any city's green or ecological infrastructure, as important as any conventional infrastructure in a city, whether roads, sewers, or electricity. Without trees, life in cities would be difficult and unpleasant. They perform so many essential services, from (again) providing cooling benefits, to managing stormwater runoff, reducing air pollution, even moderating noise in cities. Despite these many benefits, cities still tend to under-invest in them, spending relatively little per capita to plant and maintain trees.

The health-enhancing aspects of trees and forests are not evenly distributed in cities, of course, but tend to reflect the systemic racism and patterns of segregation found in American cities especially. Without tree canopy many health conditions are worse—urban heat, air pollution, noise, among many others. Tree planting and tree conservation in turn represent essential tools in redressing these long-standing inequities and in bringing about more socially just cities. I will discuss in more detail later the goal and vision of "tree equity" that many cities are now seeking to bring about. Decisions about how many and where to plant trees, and about building the community institutions and capacity to grow and maintain the trees and forests in one's neighborhood are, moreover, an essential element in creating a more inclusive city. There are many ways to foster inclusivity, and to shift local decision making toward a model of co-production and co-design, but trees and forests represent an especially important opportunity.

The mental health benefits of abundant urban trees and forests are especially important. Indeed, it is hard to identify a single step in the design of cities that would have more impact than trees, and there is a growing body of research to support this. A 2018 Cigna national study found remarkably high levels of reported loneliness in the US, with nearly half of the 20,000 respondents reporting "sometimes or always feeling alone".[5] More urban trees are also at least part of the answer to society's rise in loneliness and in social isolation.

A study published in 2023 that looked at a database of more than 109,000 residents in New South Wales, Australia, concluded that dementia-related deaths and hospitalizations were significantly lower in neighborhoods with higher tree canopy cover.[6] Precisely why was not clear but the authors did look at what they call "candidate mediators," such as hypertension, diabetes, physical activity, social loneliness and sleep. These were found to explain some but not all (or most) of the effect.

It makes sense, of course, that more trees and canopy mean better sleep, and that translates into better health overall. Similarly, the more forested the neighborhood or city the more likely residents are to be outside and walking, and thus the healthier they are likely to be. The authors speculate about other possible causal mediators, including air pollution, urban heat, and noise, all likely to be lower when trees and canopy are more abundant.

US Surgeon General Vivek Murthy argues in a recent op-ed in the *New York Times* that we have become a "lonely nation" and that loneliness and social isolation ought to be considered a "top public health priority."[7] He suggests the need to develop a "national framework" to rebuild special connections, and that investing in social infrastructure is key. Although Murthy does not mention trees or nature specifically, they are a powerful kind of social infrastructure. Trees create the context for social interaction in cities—they help break down social barriers, foster trust, and create the safe and attractive spaces (shaded and natureful) where conversations can happen, neighbors can meet, children can play, and friendships can form. The more trees we have in our urban neighborhoods the more likely we are to walk and stroll outside, with many physical and mental health benefits accruing.

There are few activities that deliver more health benefits, mental and physical, than a walk in the forest. A relatively lengthy and robust body of research from Japan shows the health benefits of so-called "forest bathing," or what is called *shinrin-yoku*. Japanese researchers show compellingly that at the end of a walk in a forest stress hormone levels are lower. One receives a boost to the immune system, and the aerosols (called phytoncides) released by pine trees are even cancer-killing. The Japanese are so convinced of these benefits they have established a network of forest therapy stations around the country. The idea has taken off beyond Japan and it is now possible to become a certified forest therapy guide through the Association of Nature and Forest Therapy.[8] There is even now a global network of certified forest therapy trails.[9]

For many of us living very busy urban lives, trees offer the chance for momentary pleasure and uplift. There are glimpses of changing color, the chance to see and hear the movement of leaves or tree branches. The sight of a tree out of the window of an apartment or an office space provides a brief respite and a momentary experience of delight. Many of the experiences of trees and forests in cities are chances for what some call "microdosing," tiny experiences that deliver big effects. Many of these small experiences of trees and forests could of course be more intentional, part of the routines of our lives. Picking up and holding an acorn, touching the bark of a tree, picking up a fallen leaf and marveling over its shape and color and intricate fractal form.

Donovan et al. (2022) compared tree planting data by census tract (trees planted by the nonprofit Friends of Trees) with data on mortality in Portland Oregon. They found that the planting of trees over a thirty year period was associated with reduction in "non-accidental and cardiovascular mortality."[10] For every 12 trees planted in a census tract, they found a more than 15% reduction in deaths. This association grew larger over time, as the trees grew larger.

Trees Are Medicine

Trees and forests, then, are essential to ensuring human health and well-being, especially for those living in and near cities. There is no question that trees are medicine. Marselle et al. (2020) studied the relationship between the presence of street trees and rates of antidepressant prescriptions in the German City of Leipzig. Not surprisingly, this study of nearly 10,000 people found that the closer you live to trees the lower the rates of antidepressant use.[11] A May 2022 study published in *Environmental Health Perspectives* demonstrates similar health benefits, in this case specifically looking at sale of medications for mood disorders and cardiovascular disease in the City of Brussels, Belgium, and finding an inverse relationship. This analysis of data at the census tract level shows that "higher stem densities and higher crown volumes are both associated with lower medication sales."[12]

The study found, moreover, that larger trees and trees with larger crowns had the most impact, a unique aspect of this study, compared to others, and a conclusion that jives well with our growing sense of the importance of larger trees in cities. The study summarizes in this way: "Large tree crowns may reduce physical and mental stress more efficiently because the reduction of both heat and air pollution depends on leaf area, which is higher in large tree crowns. Psychological effects and indirect nature experiences provided by large trees, which are often old trees, may further strengthen the health impacts of these trees. Our results demonstrate that conserving large, old trees in urban environments supports not only the conservation of biodiversity but also human health."[13]

A study of thirty years of tree planting in Portland, Oregon, reached similar conclusions about their health-enhancing powers, showing a significant relationship between mortality and tree planting.[14] The more trees, and the larger these trees become, the greater the impact in reducing non-accidental deaths. For every 12 trees planted in a census tract, they found a more than 15% reduction in deaths. The association grew larger over time, as the trees grew larger. These are arguments that ought rightly to gain the attention of any public health official, or any mayor or elected city councilor.

In many cities there are remarkable numbers of larger trees and groves of trees doing this kind of work that the broader public may not appreciate. A 2016 Tree Canopy Assessment for the City of Seattle, for instance, found that there were more than 6000 large-diameter trees (trees at or above 30 inches DBH Ed: Can we add a footnote here: DBH stands for Diameter at Breast Height, a standard method of measuring tree size.), and more than 3000 tree groves (groups of at least 8 trees, each with a diameter of 12 inches or greater).[15] This information even led to the creation of a citizen group called "The Last 6000 Campaign," aimed at protecting what remains of these older trees (and discussed later in the book).

Trees Are Essential Urban Habitats

Without trees and healthy urban tree canopy many other forms of nature that we enjoy in cities would be absent, or mostly so. Birds are the most obvious example, as trees serve as an essential habitat for them. But a host of other lifeforms and an immense amount of biodiversity is supported especially in/with larger trees. A recent study of public plazas in one European city found a strong correlation between the presence of trees, and especially older trees, and the diversity and abundance of birds.[16]

Much of the therapeutic and mental health benefits are a result of these larger nature-interactions—from hearing birdsong, for instance, and the many other natural sounds of animals that inhabit trees for at least a part of their life cycle, e.g. tree frogs and flying insects of various kinds, among others.

The positive mental health benefits from trees must also be connected to their intrinsic beauty. The beauty of trees is an essential aspect of their value, though not widely discussed. For me this is an absolutely critical role that trees and forests play, and a main reason they provide such delight in our lives. Large trees are more visually complex and I often find myself focused on a particular set of branches or a unique crown shape. It is hard to imagine urban spaces, at least spaces we love and want to spend more time in, that lack trees.

Beauty and Meaning

Beauty is often understood as a highly subjective thing: an "eye of the beholder" phenomenon, and to some degree it is. But there is a strong case to be made that our attraction to nature—to trees, flowers, water—are innate and hard-wired. This is the key idea behind Biophilia: that as a species we have coevolved with nature, and that as a result we prefer certain kinds of landscapes and natural features because they have delivered evolutionary benefits for us.

Many of the trees we love the most are ones that deliver remarkable colors. In my own part of the world the joy associated with the emergence of spring is closely connected to the flowers that bloom around us: the brilliant pink and purple flowers of the redbud tree, as well as the beautiful flowers of dogwoods, magnolias, tulip poplars and yellowwoods among many others. In an evolutionary sense our response to such flowers can be understood as developing over millennia. We recognize them to be beautiful because these are cues of an abundant landscape; an environment that delivers food, water, and sustains us and helps us survive. Jeannette Haviland-Jones and her team at Rutgers University have demonstrated the power of flowers through some clever experiments in which they give random flowers to unsuspecting individuals and evaluate their facial responses. A gift of flowers (in contrast to a toy or a pen) were found to elicit the so-called Duchenne Smile or the true human smile indicating positive emotional response (and difficult to fake).[17]

Trees provide so much beauty in part because of their flowers, but this is not it alone. The larger shapes of a tree's crown, and its elaborate structure of trunk, boughs, branches, and leaves (especially as they change colors with changing seasons), are all elements of their beauty as well. The diversity and complexity of their shapes and surfaces contribute as well to their beauty.

There is some evidence that we favor certain tree shapes over others, based on the evolutionary benefits they have provided. Judith Heerwagen and others have demonstrated that we tend to prefer the aesthetics of "climbable trees" trees with boughs and spreading crowns sufficient to provide safe spaces in the event of needing to escape quickly or that provide shelter from the elements.[18] Conversely, we tend not to prefer overly tall and straight trees that do not offer those benefits. Our beginning as a species, on the southern plains of Africa—often referred to as the savannah hypothesis—is often suggested as a partial explanation for why we prefer landscapes of trees interspersed with short grasses (a precursor to the suburban lawn it is thought).

These are interesting hypotheses and propositions and may help to partially explain our emotions of beauty. EO Wilson, the originator of the biophilia hypothesis, was quick to acknowledge that our preferences and behaviors—our human nature, if you will—derives from a mix of "nurture and nature," likely more the former than the latter. This is undoubtedly true, and even for strong advocates of biophilic design and planning such as Stephen Kellert, the innate attractions we have to trees and other forms of nature will require education and cultivation. In my own case, growing up with trees all around me, and with the positive reinforcement in childhood of the value and benefits of trees by my parents, has certainly shaped my sense of beauty.

As with paintings, sculptures and other forms of art, a deeper appreciation of that art depends on training and on education, and also encouragement. Seeing the details that an untrained eye cannot the deeper intent and story behind a work of art, being open to learning about that artwork from the vantage point of others, will all play a part in what you think and feel about it. It is the same I believe when it comes to trees. The more we look at them, notice the remarkable detail and complexity, and learn about their natural history, the more we care about them and for them, and the more beautiful we will find them to be!

We have strong evidence of how important trees have been as elements of beauty and inspiration throughout history and how we have modeled human buildings and architecture after trees and forests. "There is an age-old connection between trees and architecture," say Rian and Sassone, writing in *Frontiers of Architectural Research*. They point to many examples in the past: the design of an underwater cistern of marble columns at the Basilica Cistern, in 6th-century Istanbul, "imitating a dense forest."[19] or the tree and floral decorations that adorned the so-called "dendriforms" found in the designs of Chinese and Japanese temples, and especially the ways medieval cathedrals mimicked trees and forests in their arches and vaults. Jana VandenGroot has speculated that there is an emerging "forest aesthetics" that is now finding its way into much of what we design, including high rise towers like the *Bosco Verticale*, in Milan. With the rise of mass timber structures, the presence of wood and the shapes and forms of trees have become even more present in the modern architecture of cities.

It is also likely that one important reason we enjoy looking at trees has to do with *fractals*—those self-repeating shapes in nature that characterize natural features such as rivers and coastlines and clouds, among others. Richard Taylor, the chair of the Physics department at the University of Oregon and a world authority on fractals, has studied them for many years. He coined the term "fractal fluency" to describe the way humans have evolved a visual system to effortlessly take in and process fractal images.[20] Trees, he told me, are composed of fractal shapes at their core. A fractal pattern, according to Taylor is "a pattern that repeats itself at different scales."[21] Zoom in on the branch or bough of a tree, or even further on the leaf, and you see a miniature version of the larger tree. Repetition of fractal shapes creates greater complexity and for the tree "an immense surface area within a given volume. The tree needs that to collect sunlight." Looking at trees and forests is pleasing and relaxing as well. They are visually delightful and beautiful to the eye in part because of these fractals and the fractal fluency we have developed over millennia.

There is considerable research about the important role of natural sounds in delivering the positive benefits of trees and nature. A recent study published in *People and Nature*, for example, sought to gauge, in a controlled laboratory setting, the role of sounds and soundscape in the perceived restorativeness of parks.[22] The louder and more pleasant the birdsong is believed to be, and the greater the "perceived naturalness" (really a function of the number of different bird species heard) the higher is the perceived restorativeness of the park. Parks and green spaces that lack theses kinds of natural sounds, birdsong especially, are not as likely to provide the restorative qualities we need. Without trees there will be few birds, and scant birdsong, and even abundant green space or open land will simply not do what we want and need it to do for the mental health of residents of city residents.

The overall conclusion of this study is the need to take the soundscape into consideration: "These results highlight the importance of considering the sonic environment when assessing the potential restorative benefits of an urban green space and managing the composition of natural and anthropogenic sounds within these green spaces."[23] Trees are natural sound generators, of course. In addition to the sounds of birds and tree frogs, trees are also responsible for what sound artist and advocate Bernie Krause calls the *biophony*, which includes the sounds of tree limbs moving and creaking in the wind, the subtle fluttering of leaves in a breeze. It is hard to understate the value and benefit of these sounds, and sights.

Another important insight from the *People and Nature* study is the potentially negative role of other kinds of sounds, car and traffic noise in particular. Participant responses show that car noises can "constrain" or "impair" the restorative benefits of nature, including trees and birds. This makes perfect sense, and further supports the need for cities to find ways both to reduce dependence on cars and roads, and to work to reduce the masking noises generated by this form of mobility (e.g., controlling the speed of cars, their muffler noises, encouraging or mandating quieter tires). Importantly, trees, especially when fully leafed out, have positive noise buffering qualities that will themselves help to control or moderate car and traffic noises (for instance, the distant whine of truck and highway traffic).

Everyday Arboreal Awe

The science of awe has been progressing impressively in recent years. I frequently argue that we ought to define a good city as one that makes it possible to experience many moments of awe, daily, perhaps hourly. And trees and forests are awe-inspiring in themselves but also the stage upon which many other awe-inducing events and phenomena occur, from synchronized blinking of fireflies on a summer's night to the emergence and rhythmic sounds of cicadas to the strained and powerful movements of tree branches and leaves during a thunderstorm.

According to psychologist Dacher Kelter, awe "is the feeling of being in the presence of something vast that transcends your current understanding of the world."[24] Trees and forests are not the only forms of nature that invoke awe, and nature is not the only source of awe. But urban trees and forests are an especially potent source. The vastness Keltner describes can take many different forms, including physical and temporal. The incredible size and spread of the crown of an ancient white oak is awe-inspiring and the way it has sustained and lived for hundreds of years is an equally impressive dimension of vastness.

I have had many feelings of awe around trees, or in response to trees. Several years ago a visit with my family to California's Sequoia trees left us stunned. Their size—some more than 100 feet around and more than 260 feet tall—was remarkable as well as their age. Some are 3000 years old. I have had similar feelings in response to trees closer to home that might not be quite so large or old, like the white oak near my university office that is likely to be nearly 300 years old. I often feel humble when I sit near to this tree. Its size and age exude for me a sense of permanence and deep history. I know this tree will die, but it has already lived so much longer than I have and has witnessed so much of the distant history that I know only from books and historic plaques.

I have feelings of awe just thinking about some of the older trees on my personal bucket list: the bristlecone pines of Nevada (one named Methuselah is estimated to be almost 5000 years old) and notable live oaks such as the Angel Oak near Charleston (South Carolina). Some, like the bald cypress trees of North Carolina, require a kayak or a canoe to see, and certainly part of the awe when seen is the journey taken to find and see these trees. Many are near to cities, of course. Some years ago, I sought out and found the so-called "Queens Giant," a very old tulip poplar tree, in an out of the way site in the borough Queens (thus the name), in New York City. According to the city's parks department it is likely over 360 years old. On the day I saw it, it's size and grandeur were remarkable but the setting surprising—a modest fence had been erected around it and though there was a small marker the feeling then was one of obscurity—that few New Yorker's likely knew of or had visited this tree (what one newspaper story described as "New York's Oldest Resident"[25]). I would like to imagine a time when the many older and wondrous trees in cities become a keen focus for those living there as well as those visiting; at least a part of the awareness one has as a resident of a natureful city.

Trees and forests are also the stage on which so many other natureful experiences of awe take place. It is the location where the nuthatch seemingly defies gravity, where the calls of a screech owl startle and awaken our curiosity, or where the remarkable mass of a tree and its crown spread and sway in the midst of a windy day.

One of the things that happens when we experience the emotion of awe is that we tend to think less about ourselves, and that is a good thing. According to Keltner, awe is transformative, in part by "quieting the nagging, self-critical, overbearing, status-conscious voice of our self."[26] Tree-induced awe helps us overcome the confining bounds of our narrow selves and self-interest and allows us to see and feel part of something larger. Ensuring we will always have trees around us in cities, especially older ones, holds the potential for the experiences of awe to occur regularly over the course of our daily lives—from a walk in the neighborhood, a view out the window, or time spent in a park. In cities of abundant trees and forests we are more likely to commonly experience many awe-inducing things—ants on the sidewalk, blue jays burying acorns, the many leaves in fall dancing with the wind on their trip to the ground.

There are many enchanting qualities about trees and forests, many things about them that cannot literally be seen: the remarkable root structure that for many species is even larger and more expansive than the visible crown. Or the complexity of the mycorrhizal fungi and the extensive and only recently discovered ways in which trees communicate with each other. Dwelling even briefly on the mystery and vastness of a single tree is an exercise in stoking awe.

Trees Shape Our Cities

For many of us who love trees and want to live around them it is clear how much they enhance the quality of place and the distinctness of the places we live in. In many cities the recognition of this important place-making role is even acknowledged explicitly in the taglines and the official, and unofficial, mottos and slogans we give to our communities—Raleigh, North Carolina, is the *City of Oaks*, for example, a nod to the importance of these trees in its history, or Atlanta, Georgia, calling itself the *City in a Forest of Trees*—among many others. Trees also provide a remarkable degree of verticality to cities, adding character and an immersive quality to the nature there. The trees define and provide different zones of habitat, places in the city where different species of birds, and other flora, fauna and fungi live.

There is a vertical (and more generally spatial) dimension to cities of course, but there is also a temporal dimension. Cities exist and develop over time, and often over a long period of time. We are understandably drawn to old buildings and spaces enjoying a sense of history and antiquity that deepens the connection to cities and urban neighborhoods.

In discussing "Deep Time Humility," Roman Krznaric in his book *The Good Ancestor*, provides an extensive discussion of the value of trees as "tangible reminders of deep time."[27] Trees, especially older ones, help serve as a bridge between generations. Conservation of historic buildings and ancient landscapes (not specifically mentioned) would seem to serve a similar function.

James Canton in his book *The Oak Papers*, writes compellingly of his personal efforts to get to know one ancient oak, the Honeywood Oak, on an estate in southeast England. He visited and sat under this tree almost every day and experienced firsthand the remarkable life it supported. It also helped provoke in him a sense of deeper time, a connection to the past. "Ancient oaks hold a powerful sense of longevity," he says. "The sense of security, the sense of attachment to a place across time, enchants us. We are drawn to old oaks. You can stand beneath a grand oak and know that your more distant ancestors did so too. Oaks hold onto the memories of older generations. By touching the skin of the oak, it is possible to feel some tentative trace of those that have gone before."[28]

"Those trees act as conduits to connect us with those who have gone before. Beloved parents, aunts, uncles, grandparents of earlier generations have also touched the ribbed bark of the tree where now living hands brush the same coarse skin. There is a sense of the corporeal connection and so a powerful feeling of remembrance in the simple act of touching an old oak. It is a bodily remembrance through time."[29] It is shocking how casually we often cut down such trees, though there is now a greater understanding of the immense amount of carbon sequestered in them (and a greater awareness of the feebleness of efforts to plant saplings in their place!

Trees are time-keepers, then, and help us to tell and remember the events and people that take place there (Figure 1.1). To the extent that we give priority to protecting older, long-lived trees in our midst, allowing them to grow (and die), they provide a kind of living history essential for preserving the collective memory of cities and neighborhoods.

Trees and forests play an essential role in shaping the character of the cities in which we live. Without trees and forests our cities and urban neighborhoods would seem to us lifeless and without the vibrancy

FIGURE 1.1 Old Trees in Cities Connect Us to the Deeper Past. Photo Credit: Anneke Bastiaan.

aliveness that trees and the many lifeforms they support will bring. And of course, in many ways trees bring beauty and delight into our daily lives; we benefit immensely and hourly from seeing and experiencing the trees and forests around us.

Tree Civics: Learning to Be Good Tree Citizens

Trees also play an important role in helping to create and maintain democracy. This may seem like a bold and unusual claim, but I believe it is true. Without the presence of trees public spaces of all kinds—public plazas, parks, streets and sidewalks—become less inhabitable and more difficult to spend any amount of time in. Protests, peaceful assemblies, parades, and political expressions of all types and shapes become more unpleasant and difficult.

There are democratic implications of another kind. Actions to protect trees have themselves become exercises in political expression. Politicians and elected officials often fail to fully appreciate the extent of the care and love people have for trees at their own peril. At time when it seems impossible to personally do anything that might make a difference, the protection of an ancient tree, or the planting of new trees, represent a tangible and significant steps toward making a difference. Actions on behalf of trees have an intertemporal quality and meaning; personal and political actions both deliver immediate benefits but also honor both the past and a commitment to the future.

As we will see in the chapters that follow, efforts to protect and conserve trees, even to study and monitor them, often require a collective undertaking and the work of many volunteers in a community. Developing New York City's city tree map, for instance, required the volunteer labor of several thousand citizens.

Protecting existing older trees on one's private lot in turn becomes a civic act of contributing to the larger public good and to the welfare of the broader neighborhood and city in which one lives. Mostly today, how you decide to use or treat the trees on your parcel (and in many cities 60% or more of the trees can be found on private property) is understood as a private matter, not one that you to think of the impact of your action on your neighbors and the larger public. But these so-called private decisions do impact the larger good—individually and cumulatively these affect how hot a neighborhood or city is, how walkable it is, and how full of birds and wildlife and thus how delightful and uplifting it will be. Saving that tree, or trees, stewarding over them and caring for them is understood as a form of commitment to the larger community of which one is a part.

Equally true trees and tree conservation represent immediate and tangible things that individuals (and groups) can do to make a difference and to address our pressing environmental challenges. The threats facing our global environment are immense, and for the most part we see them as beyond our control. What can we realistically do to curtail Amazonian deforestation or reduce climate change? It is hard not to be discouraged when we look at the global trends. The earth's environment is quickly unraveling, as massive habitat loss, excessive overconsumption, and the impacts of climate change are ever apparent. The IPBES Global Assessment has estimated that about one million species are at risk of extinction, as we enter a period some have called the *Sixth Mass Extinction*.[30]

Conserving trees and forests are an important part of the necessary response to this global biodiversity crisis. And urban trees and forests are a partial anecdote to the sense of hopelessness that sometimes takes over, the sense that there is nothing that can be done. There is of course, and it may be as rewarding as planting trees in the backyard or a corner vacant lot. Or standing up for an ancient (and high carbon storing!) tree in one's neighborhood. These are real ways to make a difference and meaningful and significant responses to the problems we face today. Cities can and must operate at a larger scale, must become leaders in global conservation, to be sure, and I will discuss some of the ways they can do this later in the book. But caring for and about trees presents an unusual opportunity to re-earth our perspectives and can serve as an essential portal for other forms of political activism and good work.

Ironically, we are rapidly losing nature at precisely the time we are learning more about it. This is especially true when it comes to trees and forests. There is little doubt that having trees around us in cities helps us to be better humans and to care more about each other. These "civilizing" effects of trees and nature more generally are sometimes underappreciated. There is considerable evidence that trees along streets help to calm and de-stress drivers, in turn reducing accidents and road rage.[31] As the previously mentioned studies on awe suggest, exposure to nature has the benefit of helping to dampen our self-centered and self-interested tendencies, and to see ourselves as profoundly connected to others and part of a larger world.

A number of studies have shown that in the presence of nature we are more likely to be generous, to exhibit prosocial attitudes and behavior, to be more cooperative and think longer term.[32] Trees are a big part of this, I believe.

Understanding Trees as Members of Communities of Life

With new research showing the remarkable and, to many, surprising powers of communication of trees and plants, sharing nutrients, sending warning signals, generally helping each other (and in Suzanne Simard's case the special role of so-called "mother trees" who are even able to recognize and help their offspring), it is not hard to see why in many cities and urban neighborhoods trees, especially larger older trees, are often seen as members of the community. They are friends and essential community members—living creatures that occupy important and prominent spaces in the neighborhoods and in places residents see and experience them on a daily basis. And for all the reasons discussed earlier they provide essential services to the neighborhoods, including healing and comfort in stressful times.

So, no wonder we often give them names and treat them as family. The recent example of a large western red cedar, measuring 44.5 inches in diameter, slated to be cut down, in the Seattle neighborhood of Seward Park is a case in point. The neighborhood sprang into action, after (accidentally) learning that an application had been submitted to cut the tree down (it qualified as an "exceptional tree" under Seattle's tree protection code). The neighbors organized and convinced the property owner to save it. They affectionately named this tree as "May" (after South Mayflower Street) and speak about it as if a person. One of the key advocates was 91-year-old resident Mordo De Jaen, who believes the tree is older than he is. He has walked around the neighborhood speaking with other residents who express a similar affection for this very large tree.

"As I went around and talked to the neighbors, there was not a single neighbor that said, 'I don't care if the tree is cut down,'" says Caryn Swan Jamero, another neighbor, quoted in the local news story about the conflict over the tree.[33]

In Native American culture trees are kin, family and community members, often referred to as "our standing people." Robin Wall Kimmerer, especially in her important book *Braiding Sweetgrass* speaks eloquently of the need for "a kinship with the world,"[34] and for a language that speaks of trees and animals and nature not as objects, but as living persons. "Naming is the beginning of justice," she says.[35]

Giving trees names may seem to some a wrongheaded step in the direction of anthropomorphizing trees. But as we increasingly recognize them as individual living creatures, this seems appropriate and a tangible expression of how we care for and love these precious cohabitants. Perhaps this approach will not work everywhere, A 2016 study of Seattle's urban forest estimates there are more than 6000 old growth trees.[36] All are worthy of naming, but there are limits to this approach. If you are lucky enough to be living on a street or in close proximity to one or more of these trees, naming does not of course take the place of legal protection, though it is a helpful precursor. And how do we go about protecting the larger urban forest, of which these individual trees (as we are) are a part? We may want to consider naming the larger forest as well.

The point is that trees and forests are increasingly viewed, as they should be, as living entities, holding intrinsic worth and deserving of moral consideration and legal protection. This is the latest set of arguments to add to why we should protect and preserve trees in cities—not only (or primarily) because we benefit from them, and which we undeniably do, but also because they are living beings that have intrinsic moral worth, or inherent value, irrespective of the utilitarian value they may have for humans. It is an expansion of our notion of citizenship—that we acknowledge the trees themselves as members of our biotic community. And it is about changing our legal system and legal projections to acknowledge that trees and forests have or ought to have rights. We will especially explore this idea of giving trees and forests legal personhood later in the book. The good news is that we are beginning to see this legal and moral shift happening in cities and more movement in these directions is inevitable and positive.

Despite these many benefits and the profound ways that trees and forests shape our cities and our lives, they face many threats today. Some of these involve building and development projects that ignore, or worse, actively destroy trees (Figure 1.2). Trees and forests still receive less attention and fewer resources than they deserve. In many cities, even in our largest, richest cities, trees and forests receive a relatively small share of the budgetary pie. In New York City, for example, the latest proposed

FIGURE 1.2 Loss of Trees from Urban Development (a home renovation in Charlottesville, VA). Photo Credit: Tim Beatley.

budget will likely see a reduction rather than an increase in the funds and staffing available to protect and care for existing trees and expand the urban canopy.[37] This despite the newest evidence about the important ecological benefits that trees provide for citizens of this city.

Trees often just seem to be in the way of urban development projects and infrastructure in growing cities. This is true it seems in cities all around the world. Trees are being lost in New Delhi and many other cities to road expansion. Several years ago an ancient banksia forest was threatened by a highway expansion near Perth, Western Australia. The community came together to stop the highway (but not before much of the forest had been lost). The prevailing view in many cities is that the loss of trees and forests is an inevitable cost of modernizing.

In many cities (perhaps most) trees and forests do not have the strong legal protection they need and deserve. It took a lawsuit to prevent the City of Los Angeles from cutting down some 12,000 trees in order to repair damaged sidewalks.[38] Trees, especially older trees, need not be cut down—there are better ways to create and maintain walkable spaces in a city and accommodate trees and sidewalks together (and a perfectly smooth sidewalk is useless if it bakes like an oven without the shade and cooling of trees).

A big part of the problem is that trees and forests are viewed especially in cities, as expendable. They hold a status something akin to street furniture: sometimes appreciated for the benefits they provide but not understood, especially larger older trees, as irreplaceable or sacred. And certainly not as vital living ecosystems that are essential to the quality of life. If a tree needs to be cut down, to make room for a new house or development, it can be easily replaced by planting seedlings or younger trees somewhere else. This attitude often prevails at the city policy of operational level, even though the reality is often much different at the level of the street or neighborhood, where trees are beloved and not thought of as "expendable" or "tradable" at all.

While there has been a growing awareness of the many ecological and other services provided by trees in cities, the full extent of the benefits provided by trees and forests is rarely acknowledged, or rarely reflected in city priorities and policy. As Jim Davis, co-founder of the Seattle tree advocacy group *The Last 6000 Campaign* told me, trees are "probably one of the city's most important public health infrastructures."[39] And the evidence of the health, and especially mental health benefits of trees, is compelling but not widely discussed among urban decisionmakers.

In many cities the loss of trees and decline in canopy happens each year in many small ways that cumulatively reduce the urban forest. The loss of a large tree here, or a few trees cleared from a vacant lot there. This is a process of nibbling away that is sometimes hard to discern or recognize. Part of the challenge for cities is to keep track of these gradual but cumulatively significant losses over time.

The good news is that there are growing signs of pushback. In many cities citizens and advocacy organizations have organized to actively oppose projects that lead to loss of trees and to advocate for greater funding and priority given to trees and forest conservation. And there are many good examples of city plans, codes and initiatives aimed at protecting trees and growing the urban canopy. A major purpose of the present book is to catalog and share these best practices and inspiring urban work so that cities can see what is possible and can learn from each other.

Thanks to new research we now have an even deeper level of knowledge of trees and forests and even more reason to be awed and to care deeply about them. I am hopeful that we will and that cities everywhere will redouble their efforts to protect and expand their urban forests. Every resident of a city is entitled to live and flourish under the sheltering beauty of an urban canopy.

Conclusions

There are many good reasons why cities can and must prioritize trees and forests and elevate their status. As cities become ever more hotter they will literally become uninhabitable without the cooling benefits provided by trees and forests. There are many other ecological services and benefits provided of course: they moderate air and water pollutants and help to retain and manage stormwater. A heavily canopied city will not be optional but absolutely essential to maintaining a livable and healthy city.

There are, moreover, so many ways in which trees and forests help to improve our lives and enhance the quality of these lives in cities. Trees provide immense pleasure and enjoyment, and it is hard to overstate how dull and sensorially stultifying urban and suburban environments would be without trees and forests. They enliven and animate cities. They also create essential habitat for many other nonhuman species, such as birds. Without trees there would be few birds and little birdsong to uplift and inspire.

The research about the benefits of trees and forests has been growing at a remarkable clip in the last decade. Especially apparent are the mental health benefits provided by trees. Many of these benefits are quite direct and fairly immediate—think of the relaxing and anxiety-dampening effects of watching a tree sway or listening to the birds singing and flitting from branch to branch. Trees in cities provide moments of awe and respite. The more trees and canopy a neighborhood possess the more likely residents are to be outside and the more likely they are to engage in physical exercise, with its own abundant health benefits. We know that trees and shade in cities provide spaces to socialize and help to pull us together. Trees and forests, like many other forms of nature in cities, tend to have civilizing effects. They help us to break out of our focus on narrow self-interest and to see ourselves as a part of something larger.

Notes

1 Robert Cazzola Gatti, et al., "The Number of Trees on Earth," *PNAS*, Vol. 19, No. 6, 2022, found here:https://www.pnas.org/doi/pdf/10.1073/pnas.2115329119, accessed May 5, 2023.
2 David J. Mildrexler et al., "Large Trees Dominate Carbon Storage in Forests East of the Cascade Crest in the United States Pacific Northwest," *Frontiers in Forests and Global Change*, November 2020, Vol. 3, found here: https://www.frontiersin.org/articles/10.3389/ffgc.2020.594274/full, accessed May 20, 2022
3 Texas Trees Foundation, Urban Heat Island Management Study, 2017, found here: https://www.texastrees.org/wp-content/uploads/2019/06/Urban-Heat-Island-Study-August-2017.pdf, accessed May 22, 2023.
4 Lungman, Tamara, et al., "Cooling Cities Through Urban Green Infrastructure: A Health Impact Assessment of European Cities," *The Lancet*, January 31, 2023, accessed May 22, 2023.
5 Cigna, "New Cigna Study Reveals Loneliness at Epidemic Levels in America," June 2018, found here: https://www.cignabigpicture.com/issues/june-2018/new-cigna-study-reveals-loneliness-at-epidemic-levels-in-america/, accessed May 30, 2023.
6 See Thomas Astell-Burt, Michael A. Navakatikyan, and Xiaoqi Feng, "Why Might Urban Tree Canopy Reduce Dementia Risk?" *Health & Place*, Vol. 82, July 23, 2023, found here: Urban green space, tree canopy and 11-year risk of dementia in a cohort of 109,688 Australians - ScienceDirect, accessed May 15, 2023.
7 Vivek Murthy, "Surgeon General: We Have Become a Lonely Nation. Its Time to Fix That," *New York Times*, April 30, 2023, found here: Opinion | Loneliness Is an Epidemic in America, Writes the Surgeon General - The New York Times (nytimes.com), accessed May 10, 2023.
8 See the Association of Nature and Forestry Therapy, https://www.natureandforesttherapy.earth/
9 More information about certified trails can be found here: https://www.natureandforesttherapy.earth/certified-trails
10 Geoffrey H. Donovan et al., "The Association between Tree Planting and Mortality: A Natural Experiment and Cost-Benefit Analysis," *Environment International*, Vol. 170, December 2022.
11 Marselle, Melissa R. et al., 2020, "Urban Street Biodiversity and Antidepressant Prescriptions," *Scientific Reports*, December, found here: https://www.nature.com/articles/s41598-020-79924-5, accessed June 16, 2022.
12 Dengkai Chi et al., "Residential Exposure to Urban Trees and Medication Sales for Mood Disorders and Cardiovascular Disease in Brussels, Belgium: An Ecological Study," *Environmental Health Perspectives*, Vol. 130, No. 5, May 11, 2022, found here: https://ehp.niehs.nih.gov/doi/10.1289/EHP9924, accessed June 4, 2022.

13 Ibid.
14 Geoffrey H. Donovan, et al., "The Association between Tree Planting and Mortality: A Natural Experiment and Cost-Benefit Analysis," *Environment International*, Volume 170, December 2022, found here: The association between tree planting and mortality: A natural experiment and cost-benefit analysis - ScienceDirect, accessed May 10, 2023.
15 "2016 Seattle Tree Canopy Assessment," found here: https://www.seattle.gov/documents/Departments/Trees/Mangement/Canopy/Seattle2016CCAFinalReportFINAL.pdf, accessed July 28, 2022.
16 Maximilian Muhlbauer et al., "A Green Design of City Squares Increases Abundance and Diversity of Birds," *Basic and Applied Ecology*, April 2021, found here: file:///C:/Users/tb6d/Downloads/Mhlbauer_et_al_2021_urban-birds-squares.pdf
17 E.g. see Jeannette Haviland-Jones, Holly Hale Rosario, Patricia Wilson, and Terry R. McGuire, "An Environmental Approach to Positive Emotion: Flowers," *Evolutionary Psychology*, January-December 2005, found here: An Environmental Approach to Positive Emotion: Flowers - Jeannette Haviland-Jones, Holly Hale Rosario, Patricia Wilson, Terry R. McGuire, 2005 (sagepub.com), accessed May 10, 2023.
18 See "Resiliency in Healthcare: Look to Nature. A Q-and-A with Judith Heerwagen," found here: https://designgroup.us.com/expertise/resiliency-healthcare-look-nature-q-and-judith-heerwagen, accessed April 7, 2023.
19 Iasef Md Rian and Mario Sassone, "Tree-Inspired Dendriforms and Fractal-Like Branching Structures in Architecture: A Brief Historical Overview," *Frontiers of Architectural Research*, Vol. 3 (298–323), March 2014.
20 More about Fractal Fluency here: "Richard Taylor: Physics and The Art of Fractal Fluency In Nature," November 27, 2020, found here: https://www.buzzsprout.com/1074598/650974
21 Interview with Richard Taylor, University of Oregon, 2019.
22 Konrad Uebel, et al., "Urban Green Space Soundscapes and Their Perceived Restorativeness," *People and Nature*, May 24, 2021, found here: https://besjournals.onlinelibrary.wiley.com/doi/full/10.1002/pan3.10215, accessed June 16, 2022.
23 Ibid.
24 Dacher Keltner, *Awe: The New Science of Everyday Wonder and How It Can Transform Your Life,* New York: Penguin Press, 2023, p.7.
25 Angi Gonzalez, "Meet the Queens Giant: New York City's Oldest Resident," June 17, 2019, found here: https://www.ny1.com/nyc/all-boroughs/news/2019/06/17/meet-the-queens-giant–the-oldest-tree-in-all-of-new-york-city, accessed April 10, 2023.
26 Dacher Keltner, *Awe: The New Science of Everyday Wonder and How It Can Transform Your Life*, Penguin Press, 2023.
27 Roman Krznaric, *The Good Ancestor: A Radical Prescription for Long-Term Thinking*, The Experiment, 2020.
28 James Canton, *The Oak Papers,* Black Inc, 2020, pp.4–5.
29 Ibid, p.13.
30 See Elizabeth Kolbert, *The Sixth Extinction: An Unnatural History,* New York: Henry Holt and Company, 2014.
31 For an excellent review of the literature see Kathy Wolf, "Safe Streets - A Literature Review," *Green Cities: Good Health*, College of the Environment, University of Washington, 2010, found here: https://depts.washington.edu/hhwb/Thm_SafeStreets.html

32 For a review of some of this literature, see Beatley, Handbook of Biophilic Design and Planning, Washington, DC: Island Press, 2017.

33 Agueda Pacheco Flores, "Seward Park Neighbors Come Together to Save an 'Exceptional' Tree," March 10, 2022, *South Seattle Emerald,* found here: https://southseattleemerald.com/2022/03/10/seward-park-neighbors-come-together-to-save-an-exceptional-tree/, accessed August 1, 2022.

34 Robin Wall Kimmerer, *Braiding Sweetgrass: Indigenous Wisdom, Scientific Knowledge and the Teachings of Plants,* Milkweed Editions, 2025, p.21

35 Robin Wall Kimmerer, "Speaking of Nature," *Orion,* March/April, 2017, p.16.

36 See Seattle 2016 LiDAR Canopy Cover Assessment, May 8, 2017, found here: https://s3.documentcloud.org/documents/6207983/2016SeattleLiDARCanopy CoverWebinarFINAL050817.pdf, accessed August 1, 2022.

37 Mariana Simoes, "Mayor Adams Promised 20,000 Trees a Year. But Budget Cuts Threaten Progress," *City Limits,* February 1, 2023, found here: https://citylimits.org/2023/02/01/mayor-adams-promised-20000-trees-a-year-but-budget-cuts-threaten-progress/, accessed May 22, 2023.

38 Grace Tookey, "LA Wanted to Cut Down 12,000 Trees. Advocates Stood in the Way," LA Times, Feb 14, 2023, found here: https://www.latimes.com/california/story/2023-02-14/trees-saved-sidewalk-revamp-los-angeles, accessed May 22, 2023.

39 Interview with Barbara Bernard, Jim Davis, Jessica Dixon, and Sandy Shettler, Campaign for The Last 6000, August 8, 2022.

2

TREE CITY VISIONS AND ASPIRATIONS

There is a growing recognition of the essential importance of nature especially in cities. From the beginning of the global COVID pandemic in 2020, many residents of cities began to reconnect to birds, trees and nature, and increasingly understood the therapeutic and stress-reducing benefits of parks and greenspaces. But well before the pandemic there have been efforts to prioritize the role of nature in cities. The idea of *biophilia*—our innate connection to nature and literally a love of life and living systems—has emerged as a compelling framework and vision for the future of cities.

Kellert describes biophilia as an "inherent inclination to affiliate with the natural world." It was to him a "birthright" and a "biological urge," but not something that is automatically expressed. Rather it is an inclination that also requires learning and cultivating.[1]

The vision of Biophilic Cities (and the forming of a global network of cities) began to take shape in 2013, and the idea of "nature based solutions" (NBS) to urban problems has emerged as an important set of parallel ideas. The key notion here is that we need and want nature in our increasingly urban lives. All of the emerging evidence is that nature is not something optional, but rather is essential to leading a happy, healthy and meaningful life. We need cities that provide for the flourishing of both human and nonhuman life and trees and forests become a key element in this vision.

Biophilic Cities and the Vision of Immersive Nature

The vision of biophilic cities is one that shifts our notion of cities from mostly hard-surface buildings and roads to something more like a natural

DOI: 10.4324/9781003377344-2

ecosystem (which in fact it is). Nature is not just restricted to designated places in the city—most typically parks—but can be found throughout. Indeed, it is a vision that seeks not just an occasional trip or visit to nature, but an experience of immersive nature, where one is living in a city but also embedded in the natural world. The vision is one that hopes to help us break out of the mental and physical separation from the natural world to a place where we feel deeply a part of nature.

As Stephen Kellert has noted, "too little contact with the natural world and our biophilic values atrophy."[2] Our disconnect from nature is not surprising perhaps given how recent human history is of cities and buildings that isolate and separate us from nature. In many places we live in air conditioned homes, work in offices with little natural light and with windows that do not open, we return each day to sterile neighborhoods with few trees and limited biodiversity. For an increasing number of us living in cities we seek a different life and a different kind of city—a biophilic city with nature and trees all around us.

Other elements of the vision include the idea of temporal immersion—that exposure to trees and nature should happen during every phase of life, starting at a young age (every school should be natureful and biophilic), and should extend into one's elderhood. There must be a fair distribution of trees and nature—as every resident is entitled to contact with nature. It is a vision that understands the need to at once restore ecological systems and to design buildings that include nature and contribute meaningfully to the biophilia and biodiversity of the city.

The city state of Singapore has emerged as a global leader and exemplar of what such an urban vision might look like: impressively as the city has grown significantly in population it has seen an expansion (not a diminution) of its natural and green cover. Not all of this is trees and forests of course, and Singapore has been a pioneer especially in the form of "vertical nature" finding creative ways to integrate nature into the highrise structures that accommodate so much of its growth. But it is also trying to bring back its tropical forest, and to reestablish even its grandest and tallest trees. And everywhere one travels in this city, it seems, there is a sheltering tree canopy. As Lena Chan, of the National Parks Board, or NParks, has explained, the city's landscape strategy embraces an ecologically complex approach of a multilayered urban forest, recognizing that native birds and other wildlife depend on these different layers.

Trees and urban forests must necessarily be a key element in advancing this biophilic vision of immersive nature. There are many examples of cities around the US and the world that have set ambitious tree canopy and tree planting targets. In Washington, DC, the current tree canopy coverage is 38%, with a goal of reaching 40% (contained in the city's Sustainable DC plan) by the year 2030. Even more ambitious, the cities of

TABLE 2.1 Tree Canopy Existing and Targets in Selected Cities

	Current Canopy	Target Canopy
Atlanta	46.5%	50%
Chicago	16%	
Norfolk	25.8%	30%
Pittsburgh	42%	60%
Portland	29.8%	33.3%
Richmond	42%	60%
Toronto	28–31%	40%
Washington	38%	40% (Potential 55%)
Wellington	30.6%	

Pittsburgh, PA, and Richmond, VA, had both set the goal of reaching a forest canopy of 60% (Table 2.1).

Few large cities in the US approach that level but some have. Notable as a larger city with an impressive existing canopy, the City of Atlanta, Georgia's has an impressive current canopy of a little over 46%. This number has been higher, closer historically to 50%, but the city continues to gradually lose trees, often because of larger homes replacing smaller homes.[3]

More trees and nature can address a variety of urban problems cities are facing of course, including heat. The Colombian city of Medellin created a series of ecological corridors, 20 kilometers in total, that brought trees and greenery back to the center of the city. Among other things the birds returned, and residents heard birdsong where it had been gone for years. City officials report that the trees and green corridors have also reduced the temperature here by 2 degrees Celsius (3.6 F).

There can be little doubt that protecting the forests and trees that exist in cities, and growing more trees where we can, is a recipe for more biodiversity and in turn the joy and delight we get from experiencing this life in cities. In many cities there are remnant forests and smaller forest patches. Research increasingly shows the important ecological and habitat value they provide, especially for migrating birds. A recent analysis of migratory bird "stopover hotspots" confirmed the value of these small forests. These are places, the authors note, "where birds can briefly rest and replenish their energy stores while moving through a generally inhospitable landscape."[4]

Cities have the ability to protect and preserve even sizable remnant forests with big impacts. In Pittsburgh, a pair of bald eagles has been nesting and raising young—a first in 150 years within the boundaries of the city. The nesting site is in the heavily forested Hays Woods, now the city's most recent park. You might say it is the "park saved by the eagles." For many years this land has been owned by private developers and there was every assumption it would likely be developed at some point. The eagles changed all that, as did the intense interest in seeing the eagles (and there is as well a very popular webcam that many watch of the nesting eagles!)

It is difficult to imagine a dense city that could support nesting eagles without trees and forests, and it is hard to overstate the value of an urban existence where one has the daily chance to glimpse such a majestic creature.

And for many cities trees and forests have played an important role in the history of their cities and are essential in defining their communities and defining a strong sense of place. This is certainly true for Atlanta, which has been known as the "City in a Forest." Raleigh, North Carolina, is another example, known as the *City of Oaks*, a reference to the original floral qualities of the 1000 acres purchased to establish the city. It is an historical reference that still holds importance and relevance today. In March 2022 I visited many sites in that city where grand oak trees loom large, including two of the remaining historic public squares dating to the city's original 1792 town plan, prepared by William Christmas, the city's founder.[5] In this original plan there are five prominent squares—a larger center square with four smaller squares. Several of these historic oak-infused squares exist today and continue to contribute to this city's sense of place.

In some cases the very origin of a city's name is found in trees or forests. Palo Alto, California, in the San Francisco Bay Area, literally meaning "tall stick" is named after the city's 1,000-year-old redwood tree. Herculean efforts have taken place over the years to ensure the tree's survival (and there is now a plan to electrify the rail line that runs adjacent to the tree, so as to reduce the diesel pollution threatening the tree today).[6] Older trees like this one become temporal connections to a city's past, as well as help to define its future.

Cleveland, Ohio, is another city that marks its beginnings to the older trees that still exist in parks and around the city. As in Raleigh the trees provide a deep connection to the city's past. This city was formed in 1796 and the older trees that existed at this time—there are more than 270—are known as "Moses Cleaveland trees" (a reference to the city's founder). A local history group, the Early Settlers Association, has taken charge of keeping a running list of these trees and occasionally measuring and keeping track of them.[7] In these ways trees become an essential measure of a city's history, marking time, connecting a city to the people and events defining its past. A sense of antiquity and history is often a prized element in what makes a good city and trees contribute mightily to this.

Canopy Cities

The vision of a city of trees, or a city in a forest, is one reflected in the mottos or tag lines for many other cities. I think it is so prevalent in large part because it captures something we feel and want deeply—to live within nature, not separate from it. We want it all around us, and imagining a city embedded in a forest is a compelling and highly attractive vision.

It is easy to underestimate the value and importance of these kinds of urban mottos. They form a collective image—a statement often of what is special about a city and embedded in them a statement about and aspiration for the future. Changing a city's motto, especially without much public input, can be controversial, as the effort recently to change the city motto for Sacramento, California's capital city. It has been for many years known as "the city of trees" and residents there have been reminded of this motto and by a prominent water tower with the words written on its side. A mostly unvetted effort resulted in painting over these words with a new motto: "America's farm to fork capital." The result has been a public outcry, and proposals to re-paint the earlier motto or at least add it alongside the new one. As one op-ed piece argues:

> For decades, 'City of Trees' has given Sacramento—the perennial cowtown, often overshadowed by our ritzy Bay neighbors—a real sense of identity that we could own. It provokes memories all Sacramentans surely share of seeking relief from our blistering summers under the canopy of trees lining our streets and parks.[8]

In Milan, Italy, the construction of the innovative twin forested residential towers, called Bosco Verticale, the design of architect Stefano Boeri, has helped to stoke the urban imagination about where trees can and ought to be found in cities. It has become in a short time an iconic image of the possibilities of "nature- and land-sharing" in cities, how trees, greenery and nature can be designed into virtually all of the spaces in a city. Sitting even on the thirtieth story of a tower, one can feel like you are living in the middle of a forest.

Images of *Bosco Verticale*—photographs, and renderings before construction—have graced many a book and magazine cover and have become synonymous with this biophilic urbanism in many ways. Boeri notes how it has become a symbol of a new Milan: "The variations in color and shapes of the plants produce a tremendous iridescent landmark in every season and it is highly recognizable even at a distance. In just a few years this characteristic has resulted in the image of the Vertical Forest becoming a new symbol for Milan."[9]

In many cities larger older trees, especially, serve as essential elements in the character of a place, equally important if not more so than built landmarks. San Francisco, like many cities, maintains a list of heritage trees. These are generally larger and distinctive trees that are typically nominated by property owners (if on private property) and once designated by the city's Board of Supervisors they enjoy a high level of protection.[10] Many of the twenty-five designated heritage trees in San Francisco are ones that are deeply associated with the history and character of the city. They include local species like California buckeye and coast live oaks but, also some non-natives that have been in San Francisco a long time, including blue gum eucalyptus. Local tree enthusiast Mike Sullivan, author of the popular book *The Trees of San Francisco*, believes one of the most iconic trees is the Monterey cypress next to Mclaren Lodge in McLaren Park.[11] He had been asked during a podcast interview which tree in the city could be considered "the Golden Gate Bridge of trees," and that was his answer. Every city will certainly have its own unique trees and forests and they will do much to influence a sense of place and character.

Trees and forests for many communities are closely tied to aspirations of safety and security. Freetown, Sierra Leone, for example, experienced severe flooding and landslides in 2017, largely an outcome of deforestation. In response the city established the initiative "Freetown the Treetown" and set the goal of planting a million new trees. Creatively using cell phone and digital technology, the initiative has also led to a framework in which residents can receive income from planting and caring for these trees, with some of these payments coming from companies and individuals wishing to offset their carbon emissions.

Ambitious Canopy Targets

Many cities are setting ambitious specific tree targets for the future. Richmond's (Virginia) Richmond 300 Master Plan, adopted in 2017, set the goal of raising the city's overall tree canopy to 60%, from its current 42%, as well as reaching minimum canopy levels (30%) for each neighborhood.[12] This city has experienced long-standing segregation and systemic racism, including through the historic practice of redlining, resulting in far lower levels of canopy in neighborhoods of color. Part of the vision of forest urbanism must include fair and equitable access to and enjoyment of the trees in a city, and nature more generally, a topic I will take up more specifically in a later chapter.

Urban forest expert and co-founder of the Barcelona-based Nature Based Solutions Institute, Cecil Konijnendijk, has put forth a most provocative rule of thumb, and one that has already been broadly embraced as a starting point for cities. He calls its the 3–30–300 rule, and he was inspired to advance it by a rule of thumb that already existed in the forestry world—namely the 10–20–30 rule (cities should plant no more than 10% of their trees in a single species, no more than 20% in a single genus, or 30% in a single family of trees). Konijnendijk believes his rule is helpful and necessary for several reasons. It is memorable and "rememberable." And it focuses not just on the canopy level, but also the need to have access to parks and greenspaces, and also the value of experiencing trees (not from the air, a perspective most of us will not have) at ground level and likely through a window. We should be able to see trees all around us in our neighborhoods. It is hard to understate the importance of the ability to watch a tree swaying in the breeze or the animated bird life that takes place in its branches (Figure 2.1). The 3-30-300 rule has already garnered substantial support by cities around the world and has become a kind of mantra for supporter of a vision of canopy cities. In some cases 30% canopy may seem too low, but for many places it will still be an ambitious goal, and one that if reached can serve as a launching pad for even higher canopy targets. The inclusion of a goal that relates to human eye-level experiences of nature is needed and is a unique contribution to the discussion of urban targets.

The ability to see trees adds immensely to the experience of a neighborhood. Konijendijk cites a city in Denmark (Fredericksburg) that has already adopted the policy that every citizen should be able to see at least one tree for their window. This seems too low a goal, though the sight of a single tree, especially a larger older tree that may have special significance, can be a source of daily delight. As Konijnendijk says: "Seeing green from our windows helps us keep in touch with nature and its rhythms. It provides important breaks from our work and can inspire us and make us more creative."[13]

Living with trees and forests all around us in cities, ideally immersed in the urban forest, is a highly attractive vision indeed.[14]

In one of the first evaluations of the 3–30–300 rule and how easily cities might reach it, the conclusions are cautionary. Nieuwenhuijsen et al found in their sample of more than 3000 residents of Barcelona, that surrogate measures of the rule were indeed associated with improved mental health. Yet, in this dense city, few residents were able to achieve or live in an environment that satisfied all three of the rule's conditions. In fact only about 43% of the sample were able to enjoy seeing three trees from their windows. In an especially dense city like Barcelona, clearly more trees are needed. This led the researchers to conclude: "We may need to dig up asphalt and plant more trees, which will not only improve health but also reduce heat island effects and contribute to CO_2 sequestration."[15]

FIGURE 2.1 Trees from a Window. Photo credit: Tim Beatley.

Sometimes cities set tree goals that later seem unrealistic. With a current forest cover of around 26%, The City of Norfolk, Virginia, concluded that a goal of 40% was too great a reach, adjusting the target downward to 30%. But at the same time the city chose to incorporate trees into its vision in some new and interesting ways. It had already been implementing complete streets, but now has embraced a green street version of this, making sure a complete street includes trees. Future tree planting will focus on those neighborhoods needing it most, those neighborhoods with canopy of less than 30%.The city has the goal of every bus stop being shaded by trees, and ensuring that major pedestrian routes in the city (connecting parks, schools and civic spaces) would be tree-lined and shaded. Amendments to the city's *PlanNorfolk 2030* also establish minimum canopy levels for new development based on zoning districts.

Even in cities with good overall levels of tree canopy, there will be wide variation between neighborhoods and parts of the city. In less affluent neighborhoods, and neighborhoods of color that have experienced long-term segregation, tree canopy follows the historic redlining maps of the 1930s. Overcoming deep inequities in the distribution of trees and other environmental amenities is a central goal for many cities. Equitable and just nature is an essential element in the emerging vision of Forest Urbanism. We don't spend enough time in urban planning and design circles discussing the beauty and aliveness that trees bring to cities. Perhaps in remarkable contrast to the grayness and dreary environments that hard-surface concrete and asphalt and steel that many cities convey, trees bring to cities color, sound, vitality. Springtime, especially, in many eastern US cities, for example, presents the arrival of a remarkable flood of vivid colors from the flowers of native trees such as red buds, flowering dogwoods, and many flowering understory plants. In some cities, such as Pittsburgh, there have been concerted efforts to plant and replant these colorful native trees. Through the Pittsburgh Redbud Project thousands of these spectacular early blooming trees have been planted, as well as other native trees, in the city's downtown, along its three major rivers.[16] The effort has been spearheaded by the Western Pennsylvania Conservancy with some $2 million in funding from a local foundation. Over seven years nearly 4000 trees have been planted, with volunteers doing much of the planting, adding color and vibrancy and delight to these public and pedestrian spaces.

Density and Trees Together

It is possible, the chapters of this book will argue, that both urban density and abundant trees and a healthy tree canopy are possible. It is also true that careless urban growth and urbanization can lead to significant losses of trees and forest cover.

Portland, Oregon, has set the goal of achieving a citywide canopy goal of 33.3%, and it is nearly there having grown its forest significantly since 2000. Yet in the most recent canopy study, in 2021, it recorded its first decline in some time, with canopy cover dropping by nearly 1% (from 30.7% in 2015).[17] Precisely what's going on is unclear, but most observers point to the city's development pressures in recent years. Even in highly progressive Portland, where public support for tree conservation is high, it is a struggle to reach these conservation goals.

In many cities denser, central cities areas will also tend to be more dominated by impervious surfaces, and fewer trees. Ambitious targets and aspirations are nevertheless possible here as well. In downtown Atlanta, for instance, current tree canopy cover is a mere 3%. And there is little doubt that there are practical limitations to finding spots and sites to plant new trees. In 2021, a Downtown *Atlanta Urban Tree Planting Plan* was prepared that sets out the goal of achieving a canopy rate of more than 10% at the end of ten years. "The challenge of planting thousands of new trees in the Downtown district of Atlanta lies in finding opportunities within the narrow corridors squeezed between multistory buildings and busy traffic thoroughfares."[18] The plan calls for planting more than 8000 new trees in downtown.

There is no question that planting trees in denser urban environments is a greater challenge than in say low-density residential neighborhoods. But there are many new ways for integrating trees and urban density, and important trend can be seen in the design of new high-rise structures that make space for trees on rooftops, terraces, and balconies for example. It may often be possible to increase the density of the city at the same time that it becomes more forested.

Investing in a city's trees and tree canopy is another way of achieving other important elements of the vision of a city. If we want to create more walkable cities, with abundant and usable public spaces, trees will be essential for activating these spaces and, especially in the face of extreme heat, ensuring they are habitable and enjoyable.

15-Minute Forest Cities

Many cities aspire to becoming walkable cities. Especially during and following the pandemic there has been growing interest in some form of the vision of the 15-minute city where it is possible to fully function and live out one's daily life primarily through walking or bicycling, within a 15-minute travel radius. The increasingly compelling vision of a car-free city is gaining traction in cities around the world, and without the presence of trees such a future reality would be difficult to imagine.

There is growing evidence that these elements of urban vision are themselves dependent on investments in trees and canopy. Residents will have little inclination to hang out in public spaces or engage in walking in and around a city if it is a sterile, hot, and uninteresting place. Trees and nature in cities make the experience of public life and of spending time walking and sitting in public spaces a matter of enjoyment and delight.

A study by Yi Lu of thousands of residents of Hong Kong, for example, concludes that there is a definite relationship between the presence of eye-level trees and the walking behavior and walking choices of residents. As this study concludes:

Two multilevel regression analyses were conducted to examine the relationships between urban greenness and (1) the odds of walking for 24,773 public housing residents in Hong Kong, (2) total walking time of 1994 residents, while controlling for potential confounders. The results suggested that eye-level greenness was significantly related to higher odds of walking and longer walking time in both 400 m and 800 m buffers. Distance to the closest Mass Transit Rail (MTR) station was also associated with higher odds of walking. Number of shops was related to higher odds of walking in the 800 m buffer, but not in 400 m. Eye-level greenness, assessed by GSV images and deep learning techniques, can effectively estimate residents' daily exposure to urban greenness, which is in turn associated with their walking behavior.[19]

Cities want to create the conditions where residents spend more time in the public realm and trees are an essential design element, likely to be as important as any other urban design element (more I would predict) in creating the conditions for what designer's call "sticky" streets and spaces; that is, places where residents not only want to visit or walk through but also want to linger and relax. There are many factors that likely contribute to the stickiness of public spaces to be sure—public art, places to sit, food and entertainment, among many others, but few seem as essential as trees. There are many specific examples of this from cities around the US and the world: spaces with trees are spaces where residents and visitors alike want to linger, want stroll and walk, want to window-shop. From Madrid to Paris to San Francisco, trees are essential to creating enjoyable and functional public spaces.

The smaller city of Reston, Virginia, is a notable case of where a compact and walkable urban form is possible alongside trees. This iconic new town from the 1960s was conceived from the beginning by its visionary developer Robert Simon as a place that would seem more like living in a park than a conventional city. It is a community "founded on the preservation and appreciation of natural areas," including some 800 acres of woodlands. Reston has a tree canopy cover of just a little under 50%, made possible by the clustering of development into a series of 7 distinct villages (Figure 2.2). "By this means it will be possible to have

FIGURE 2.2 Trees and Forests Are Found throughout the New Town of Reston, Virginia. Photo Credit: Tim Beatley.

wooded recreational spaces adjacent to the houses," notes the 1962 Reston Master Plan, "of far greater value to children and adults alike than useless front lawns."[20] Water is also another important community feature, with bodies like Lake Anne, serving a focal point for public and civic space. It is true there are few conventional lawns here. There is extensive creek restoration underway and a popular environmental center that provides numerous environmental programs for residents. But it is the trees interspersed throughout the community and abundant wooded trails that greatly add to the feelings of connection to nature there.

One of the most important steps a city can take to support a vision of immersive nature and forest urbanism is to design and maintain a robust network of pathways, trails, and pedestrian movement corridors through the city. These become practical means to get to places (e.g., walking to work) but also opportunities to explore and learn about the nature around you, to listen to birds and other animals, to inhale scents and aerosols of trees, and to generally revel in being and moving outside.

There are many wonderful examples of cities that have invested in natureful pathways and trails that permit these kinds of experiences. These include Rio De Janeiro, Brazil, with its Trilha TransCarioca, and Wellington, NZ, where there is an extensive network of trails through and around the city. In Singapore, there are now more than 350 kilometers in its Park Connector Network.[21] These are remarkable pathways that lead residents to major parks, such the Botanical Gardens and Horticultural Park, but they are also experiences of nature in themselves. My favorite segment is the Southern Ridges, where one is able to walk along an elevated trail that moves one through the tree canopy and provides spectacular views and perspectives of the city (see Figure 2.3). There are surprises along the way—an encounter with a monitor lizard, in my case, but many birds and butterflies especially. And there are places where NParks (the National Parks Board) is attempting to bring back some of the largest native trees that existed before the city's development.[22]

A city's pedestrian network is ideally forested itself so that walking, strolling and jogging, are accompanied by the experience of being immersed in a forest. But these networks will also ideally provide access to larger forested areas in the city. In the location and design of major public buildings and facilities, such as hospitals, direct physical access to urban forests should receive priority. An example of this can be seen in the design of the Perth Children's Hospital, in Perth, Australia, where a new pedestrian and cycling bridge provides direct access to (and crossing 3 meters

FIGURE 2.3   Shown Here Is the Southern Ridges Segment of Singapore's 350-kilometer Long Park Connectors Network. Here It Is Possible to Walk through the City at Tree-Canopy Level with Remarkable Views and Close-Up Interactions with Butterflies, Birds, and Nature. Photo Credit: Tim Beatley.

above several lanes of roadway below) the wild and heavily forested Kings Park. The bridge, known as the "Kids Bridge" (the bridge is also known by the Noongar word for children, Koolangka), is a colorful rainbow, winding its way to the park.[23]

There are numerous obstacles to creating truly walkable places—for instance the well-established phenomenon that urban residents tend to overestimate the amount of time walking takes to reach destinations. Trees and a healthy canopy help to overcome these obstacles, making the walk more enticing and pleasant.

These visions are intersecting and potentially reinforcing. As more cities undertake actions to reduce auto dependence, and work to recapture some space back from the car, there in turn are opportunities to plant more trees and to expand the urban forest. One recent study in Melbourne, Australia, concludes that "redundant parking" could open up additional space in that city for nature and trees. A remarkable percentage of urban space, especially in American cities, is devoted to cars and car mobility (e.g., roadways, surface parking). Shifting mobility even to a small degree away from cars will help create the conditions for higher tree canopies and more tree-immersive experiences in cities.

Creating a city of abundant and interesting pathways and opportunities to spend time outside is a primary strategy for overcoming the (epidemic of) social isolation. We need cities to get lost in; Rebecca Solnit famously writes of the need to get lost, in places where, as she eloquently says, "you constantly acquire moments of arrival, moments of realization, moments of discovery. The wind blows your hair back and you are greeted by what you have never seen before."[24] Creating spaces where it is possible to see and interact with others, spaces for intergenerational contact and play, providing outdoor settings in which residents can mutually engage in and activities that inspire fascination and awe will help to make cities healthier and residents happier.

We want to create cities that citizens are able to explore, to be curious about, to find some level of adventure even. In the best cities this is made easy by investments in pathways and walking and hiking infrastructure, and often this takes residents through forest groves and patches.

David Roberts, famous mountain climber and adventure writer, in his memoir *Limits of the Known*, speaks of the desire for adventure as something innate. "I'd like to think that what we call the zest for adventure lies dormant in all human beings, numbed by the creature comforts of home and the tedium of the job. I'd like to believe that wanderlust is encoded in our DNA."[25] Whether we have such an instinct is debatable, but there is little doubt that by designing our cities for walking and hiking, especially in ways and in areas where one can experience trees and forests city life is profoundly improved.

Cities of Wildness and Awe

Increasingly the vision of future cities we want to advance is one in which there is also *wildness*. Cities will never be places of wilderness in the usual meaning of this term: remote places that require extensive travel to get to, where there are few and any other humans, only to be visited briefly; spaces and places understood to be pristine, where the evidence of humans is limited or absent entirely.

We know now that there are few places that are pristine. That said, humans must resist the view of seeing themselves distinct and separate from nature. Wildness, and wild nature, can and does exist in and near to cities. And cities can design and plan for wild spaces; It is possible to make room for wildness and experiences of wildness in cities. Urban forests, small and large, are especially promising venues and spaces for urban wildness.

Urban ecologist Marielle Anzelone has spent much of her life countering the popular impression that there is little wild nature in cities. Based in New York City she has adopted some unique and effective strategies for raising awareness of residents there, including pioneering the idea of popup forests, something she has undertaken to great effect several times in the city. One of Anzelone's favorite parks in New York is Inwood Hill Park, on the northern tip of Manhattan Island. Here there remains wild nature and some of the original forest ecosystems that existed when the native Lenape lived here.

She hopes cities like New York will shift their points of view away from a narrow emphasis on the climate benefits of investing in trees and on an overly simplistic (ecologically that is) view of street trees. Street trees are important to be sure, but she is especially critical of a failure to give more attention, even in a large and highly populated city like New York, to the remnant forests and more ecologically complex and biodiverse remaining lands found throughout the city.

There is the need to reimagine and redefine the street tree, she believes, and really to move beyond it if we can. Her idea is to shift from the singular emphasis on street trees to promoting something closer to smaller but diverse urban forests. This alternative vision is what she calls the *City Block Biome* (Figure 2.4). We should not settle for a tree on a street but instead "mini forests on city blocks …" there is an emphasis on biodiverse plant assemblages rather than single, stand-alone trees. We want to plant more native species of trees but also "underplanting with

MARITIME MARSH FOREST_**BIOME BLOCK**

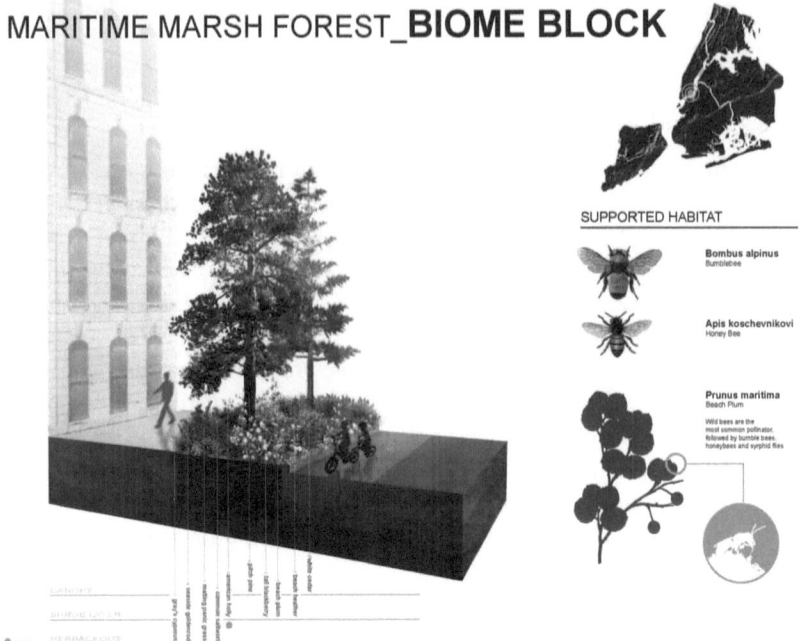

FIGURE 2.4 City Biome Block, by Marielle Anzelone. Source: Marielle Anzelone.

shrubs and wildflowers and ferns and include milkweed for Monarch butterflies." New Yorkers do not think much about Monarch butterflies, even though they have been flying over the city, she notes, for thousands of years. In short, we need more complete and complex forms of nature on our streets.

New ideas for taking up hard-surface pavement can be found in many cities such as work in San Francisco to create sidewalk gardens. Here, a special permit—for sidewalk landscaping—gives residents the power to do just that, and to insert flowers and nature that enhance the experience of walking and spending time outside in neighborhoods such as the Mission District.[26]

The City of Paris has begun planting small forests in a number of public spaces around that city, including behind an opera house, along the sides of its city hall, and on the north side of a major train station (the Gare de Lyon).[27] As well, several years ago the city began planting trees and forests in schoolyards, under a program known as OASIS, with some seventy-five schoolyards having already been planted.[28] Many of its street trees now have at their base small gardens, where understory plants and bushes have been planted (see Figure 2.5). These are wild and woolly spaces that, while not quite what Anzelone has in mind, help at least a little in shifting the model of a line of street trees toward something more vertical and potentially more biodiverse. Many of these spaces have specific names attached to them, and like allotment gardens, are planted and cared for by citizens who have taken this on.

FIGURE 2.5 A Tree Garden in Paris. Photo Credit: Tim Beatley.

It is also true that forest urban vision can extend beyond the individual city to the larger metropolitan area. Immersive nature at the neighborhood level is a key element of the vision of forest urbanism, but so also is the preservation and regeneration of larger blocks of forest. Here the distinction between land-sharing and land-sparing is relevant. A forested and dense urban neighborhood is one in which trees and forests co-occupy space with homes, offices and buildings—essentially land-sharing. As this book argues, we must be creative in identifying the ways (and places and spaces) in which we seek to grow new trees and forests in cities. It might be along streets (or in them) but also on the rooftops of buildings or the tops of highways. Land-sparing--containing urban growth tp prevent the loss of intact forests, often on the edge of cities, will also be important (more on that below).

More broadly, the vision of forest urbanism supports the notion of urban ecological regeneration. Whatever is designed or built in cities represents a chance to restore and to regenerate the biodiversity that existed earlier in time. Buildings can be biologically regenerative through rooftop meadows and forests and other ecological design elements. Every urban neighborhood and urban precinct should be judged by how many more species can be supported there over time, how many ecological functions and processes can be repaired. There are many good examples to cite. Vitoria-Gasteiz, Spain, famously converted its municipal airport into a RAMSAR-designated wetland. Curridabat, Costa Rica, through its Sweet City initiative, is planting pollinator-friendly landscaping along its streets and parks, creating biocorridors through the city.

Many of the opportunities to plant trees and to protect existing forest areas in cities will also help a city to adapt to climate change. Cities can serve an important role in creating climate "refugia," for instance, places where plants and animals might find safe harbor in the face of rising heat, flooding and other climate impacts. Many forested parks in and near to cities will represent areas of deep cool, for example, or where there is topographic diversity that will help some species adapt. And the canopy in cities will also help to connect these refugia.

A City of Many Forests

Compact and dense development patterns also present the possibility of land-sparing. Here, larger regional ecosystems and forested landscapes can be protected and left in a largely undeveloped state. Examples include extensive greenbelts in cities like Wellington, New Zealand, or Toronto, Canada, or the extensive forested land beyond the regional Urban Growth Boundary in Portland, Oregon. There is the vision of the Metropolitan

Forest in Madrid, a 65-kilometer forest belt that circles that city. These larger urban forests ideally are ecologically connected to the trees and forests in more urbanized parts of the city, and also to larger national parks and continental-scale forests and conservation lands.

Residents of a forested city in this way are surrounded by trees and forests where they live and work; these spaces of trees and density together make up the bulk of their daily dose of urban nature. But larger, more peripheral forests become places to explore and visit on weekends and holidays. In New York City there are trees and forests nearby but larger forests to be found and explored further away. New York City, for example, has purchased and protected more than 150,000 acres of land in the Catskills watershed, much of it forested, as part of its innovative effort to protect the primary source of the city's drinking water.[29] Much of this land is also available for recreational use.

Many cities today aspire to protect ecological connectivity that permits movement of birds, butterflies and other biodiversity through and beyond a city. Biological and ecological connectivity is an increasingly important element in the vision of future cities and urban forests will necessarily assume a major role.

Cities also aspire to being fair and inclusive and this applies to trees and forests as well. A vision of an equitable city is in part one in which residents have a significant say in how decisions are made and how change is managed over time. Just as important, then, in a vision of a forested city, is the direct engagement and participation of citizens and having a voice in shaping its future. This can happen in many ways, but increasingly cities provide direct opportunities for planting and caring for urban trees.

In many cities there are opportunities to learn about caring for trees. There are more than 100 Tree Tenders groups in Philadelphia, for instance, a program managed by the Philadelphia Horticultural Society (PHS). Citizens are trained in tree planting and tree care and participate directly in events in their neighborhoods. And PHS creatively utilizes online maps to help residents find tree events in the city and to identify priority planting areas.[30]

A Kinder, Compassionate City

And a city where residents are immersed in trees and nature will very likely be a kinder and more compassionate city. There is considerable evidence, already mentioned, of the humanizing and civilizing effects of trees. They help to buffer the many stresses of modern urban life, and they help us to connect deeply with others. We know from research we are more likely to be generous in the presence of nature.[31] As much of the recent evidence around the research on experiences of awe shows when we experience the immensity and vastness of nature (an ancient tree in the city, certainly, but also the vastness of an ant colony) we have a tendency to moderate our selfish, self-centered tendencies. If we are concerned, as we should be, with creating the conditions for a truly ethical city, investing in trees and working to grow a city's tree canopy, may be as important, perhaps more important, than many of the other institutions and educational initiatives that seek to create a more just, caring and compassionate city.

One of the key challenges in designing future cities is the immense and complex metabolism they constitute: urban life requires large and increasingly distant supply lines, including food, energy, water, building materials, to name a few. And there are sizable negative outputs from urban production and consumption: solid waste, air and water pollution, and greenhouse gas emissions. Rapidly reforming and redesigning this urban metabolism will take at least three steps: first, reducing the size of the throughputs, for instance by reducing the amount of energy buildings use. Then, second, doing what we can to reduce the supply lines; meaning we source more locally, rebuild our ability in and near to cities to grow at least some (if not most) of the food we need, and the energy we require. Finally, we need to move quickly in the direction of a *circular* urban metabolism, where waste and outputs are re-used as productive inputs to other activities and needs. Urban trees and forests can be important elements in re-imagining the metabolism of cities. Working forests in and around cities can be a source of sustainably-harvested wood, substituting for more distantly-sourced wood, and used for instance in local construction. Conservation of of trees and forests, as discussed in this book, can reduce the energy consumption and greenhouse gas emissions of buildings and cities. Downed trees in a city, moreover, can be recycled and reused in support of a more circular urban metabolism.

Conclusions

Crafting a vision for a city is a serious task and an important step in charting its path forward. Cities have embraced many different kinds of visions, but trees and forests appear frequently and figure prominently in them (Figure 2.6). Increasingly we are reimagining cities as not simply places where there might be bits and pieces of nature—parks to walk to and visit for example—but rather as ecosystems themselves, as places to be immersed in nature. Singapore has taken this idea the furthest—most recently conceptualizing their citystate as a "city in nature." An important part of this vision is to aspire to being a part of, rather than separate from

FIGURE 2.6   Older Trees in Northwest Washington, DC. Photo Credit: Tim Beatley.

nature. Being immersed in a forest, an urban forest, is captured in many of the stated visions of cities (Atlanta and Sacramento, for instance). Few natural elements are more essential in helping to achieve the vision of immersive nature many cities in the Biophilic Cities Network aspire to. And increasingly we want to feel and experience a kind of wild nature around us, loosening the tight reins of control and management, giving over more space to nature, aspiring to a wilder urban forest that is more than street trees planted in a line.

But there are, of course, many other important elements of a vision for future cities. The ability to radically shift away from automobiles, and toward walking and bicycling and radical proximity of the vision of a 15-minute city (or a 10-minute, or 20-minute city). We want to see dense, compact cities, and indeed these urban qualities will be necessary to bring about this more car-limited future. It is possible, I believe, to achieve a dense urban neighborhood, one that is deeply walkable and sociable, but also heavily canopied. Indeed, these elements of trees and density can and must go together. Trees and urban forests help in many important ways to advance all of these parallel and overlapping visions—more trees will make walking possible and enjoyable, for instance. Achieving an equitable and just city (more about that later) is another important part of the future vision, and the fair distribution of trees and canopy must be a part of this as well.

Notes

1 Stephen Kellert, Birthright: p.xiii.
2 Ibid, p.xiii.
3 Se Drew Kahn, "As Atlanta Grows, Its Trademark Tree Canopy Suffers," *Atlanta Journal-Constitution*, January 20, 2023, found here: https://www.ajc.com/news/as-atlanta-grows-its-trademark-tree-canopy-suffers/NM7Y6L3XUBDZTJYWPXHNIV5WC4/
4 Fengyi Guo et al., "Autumn Stopover Hotspots and Multiscale Habitat Associations of Migratory Landbirds in the Eastern United States," *PNAS*, Vol. 120, No. 3, 2023, found here: https://www.eurekalert.org/news-releases/976398#:~:text=The%20paper%2C%20%E2%80%9C%20Autumn%20stopover%20hotspots%20and%20multiscale,Jeffrey%20Buler%20and%20Jaclyn%20Smolinsky%20%28University%20of%20Delaware%29, accessed May 22, 2023.
5 "Plan of Raleigh, 1792," found here: https://digital.ncdcr.gov/digital/collection/p249901coll26/id/2800#:~:text=William%20Christmas%2C%20a%20senator%20and,of%20the%20future%20State%20House, accessed May 18, 2023.
6 Jim Robbins, "The Unlikely Survival of the 1081-Year-Old Tree That Gave Palo ALso Its Name," *New York Times*, June 26, 2021, found here: https://www.nytimes.com/2021/06/26/us/palo-alto-redwood.html, accessed May 18, 2023.
7 Grant Segall, "These Trees Grow in CLE: Groups record, track remaining pre-1796 Moses Cleaveland Trees," *Freshwater Cleveland*, October 21, 2021, found here: https://www.freshwatercleveland.com/street-level?pg=8, accessed May 22, 2023.
8 Barbara Harvey, "OPINION: Sacramento's New Water Tower Slogan Can Go Fork Itself," September 11, 2017, found here: https://statehornet.com/2017/09/opinion-sacramentos-new-water-tower-slogan-can-go-fork-itself/, accessed June 7, 2023.

9   "Vertical Forest," Boeri Architects, found here: https://www.stefanoboeriarch
    itetti.net/en/project/vertical-forest/
10 For more information see "Landmark Trees," here:
11 See Mike Sullivan, *The Trees of San Francisco*, Wilderness Press, 2013.
12 See City of Richmond, VA, *Richmond300 Master Plan*, found here: https://www.
    rva.gov/sites/default/files/2021-03/Thriving%20Environment.pdf, accessed June 3,
    2022.
13 See Cecil Konijnendijk, "The 3–30–300 Rules for Urban Forestry and Greener
    Cities," *Biophilic Cities Journal*, Vol. 4, No 2, found here: https://stat
    ic1.squarespace.com/static/5bbd32d6e66669016a6af7e2/t/
    6101ce2b17dc51553827d644/1627508274716/330300+Rule+Preprint_
    7-29-21.pdf
14 E.g. see Dave Kendal et al., "Global Patterns of Diversity in the Urban Forest:
    Is There Evidence to Support the 10/20/30 Rule?," *Urban Forestry & Urban
    Greening*, Vol. 13, No. 3, 2014, pp.411–417.
15 Ibid, p.4.
16 For more information see the Pittsburgh Redbud Project, found here: https://
    waterlandlife.org/trees/pittsburgh-redbud-project/#:~:text=The%20Inspiration
    %20to%20Plant%20Redbuds&text=This%20idea%20was%20developed
    %20by,redbud%20and%20complementary%20tree%20species, accessed April
    10, 2023.
17 See Portland Parks and Recreation, Tree Canopy Monitoring: Protocol and
    Monitoring from 2000-2020, February 2022, found here: https://www.
    portland.gov/sites/default/files/2022/tree-canopy-monitoring-2020.pdf, accessed
    June 3, 2020.
18 Atlanta Downtown, Downtown Atlanta Urban Tree Planting Plan, January
    2021, found here: https://www.atlantadowntown.com/cap/areas-of-focus/
    sustainability/downtown-atlanta-tree-planting-plan, accessed May 24, 2022.
19 Yi Lu, "The Association of Urban Greenness and Walking Behavior: Using
    Google Street View and Deep Learning Techniques to Estimate Residents'
    Exposure to Urban Greenness," *International Journal of Environmental
    Research and Public Health*, 2018, p.15.
20 *Reston Master Plan*, 1962, p.4.
21 For more about the SIngapore Park Connector Network see: https://www.
    nparks.gov.sg/gardens-parks-and-nature/park-connector-network, accessed April
    28, 2023.
22 For more about these initiatives see Timothy Beatley, *Handbook of Biophilic
    City Planning and Design*, Island Press, 2017.
23 More about the Kid's Bridge can be found here: https://pchf.org.au/the-kids-
    bridge/, accessed April 28, 2023.
24 Rebecca Solnit, *A Field Guide to Getting Lost*, Penguin Books, 2006.
25 David Roberts, *Limits of the Known*, W.W. Norton, 2019, p.279.
26 For more about this story see Beatley, *Biophilic Cities*, Island Press, 2011.
27 India Block, "Paris Plans to Go Green by Planting 'Urban Forest' around
    Architectural Landmarks," *Dezeen*, June 26, 2019, found here: https://www.
    dezeen.com/2019/06/26/paris-urban-forest-plant-trees-landmarks/, accessed June
    22, 2023.
28 As of 2022, see "Paris Oasis Schoolyard Programme, France," Climate Adapt,
    undated, found here: https://climate-adapt.eea.europa.eu/en/metadata/case-
    studies/paris-oasis-schoolyard-programme-france, accessed June 22, 2023.

29 See Committee to Review the New York Watershed Protection Program, *Review of the New York City Watershed Protection Program*, Washington, DC: The National Academies Press, 2020.

30 E.g. see "Tree Tenders Planting Opportunities," found here: PHS Tree Tenders Planting Opportunities, Philadelphia (arcgis.com), accessed May 11, 2023.

31 E.g. see Nicolas Gueguen and Jordy Stefan, "'Green Altruism': Short Immersion in Natural Green Environments and Helping Behavior," *Environment and Behavior*, Vol. 48, No. 2, 2016, pp.324–342; see also the extensive work around awe, Dacher Kelter, *Awe: The New Science of Everyday Wonder and How It Can Transform Your Life*, Penguin Press, 2023.

3

TREE CODES AND REGULATIONS

Cities have been regulating the cutting of trees for many years. Trees and urban forests, as argued here, provide many public benefits, and protecting trees is a matter of public interest. In many cities tree codes are controversial, seen by some as an infringement on personal freedoms and freedom of choice about how to use and care for trees, as well as an undue restriction on private property. But unrestricted use of private property is illusory and property rights are always subject to collective constraints. Decisions about trees and forests are no different. Each city will find a level of public control (from little to a lot) that its culture and politics can support, but strong tree codes are valuable especially in ensuring that larger, older trees are not unnecessarily or casually lost.

Many of a city's trees are located on public land or public parks, as well as in public rights of way, but some of the greatest controversy arises in situations where cities have established regulations that apply to trees on private property. In many cities much of the urban forest will be composed of trees on private land. In Boston, as reported in their new urban forest plan, this is more than 60% of the city's trees.

Nevertheless, cities have sought to regulate and protect even trees on private property in many different ways. It is an established authority that recognizes that the conservation of trees and forests serves a major public function or purpose and that conversely that the cutting or destruction of trees harms this broader public interest and undermines public health.

DOI: 10.4324/9781003377344-3

A Variety of Approaches

City ordinances vary considerably in their specifics. Most apply restrictions on the cutting of trees over a certain size, though there is wide variation in the thresholds chosen by cities and the definitions of trees that are covered by codes. Some local tree codes also apply restrictions to groves or larger groupings of trees.

A number of cities have some form of heritage tree program. San Francisco and Minneapolis, for example, have such programs. These are trees that are required to meet certain size and age criteria and are individual trees that must usually be nominated by someone in the community. In San Francisco, it is ultimately the Board of Supervisors that must vote to give a tree a heritage designation, and once that happens a special permit is required to cut down or remove such a tree.[1] In most cases these provisions apply to relatively few numbers of champion or specimen trees in a city. In the case of San Francisco there are at present only twenty-five specific trees in the city that have been designated as heritage trees.

Many cities today have established development regulations that stipulate achieving a minimum tree canopy level, often by a certain future date, say ten or twenty years in the future. Norfolk, Virginia, adopted amendments to its zoning code in the fall of 2021 to do precisely this, establishing minimum required canopy (to be achieved no later than twenty years after planting) by type of development and zoning district. For commercial, industrial and institutional zones, the minimum required canopy is 10%. It rises to as high as 20% in residential districts, and where lot sizes are 5500 square feet or larger.

In Norfolk these requirements are triggered by the need for a development permit. Developers must prepare and submit a Tree Canopy Plan, identifying trees that will be protected or planted and how the minimum canopy percentage will be met. Such canopy standards are a positive step, though in the case of developing a forested parcel, these minimums will allow quite a lot of tree loss.

Asheville, North Carolina's approach is similar, establishing minimum tree canopy standards for new development by Resource Management District (Downtown, Urban and Suburban, in turn keyed to categories used in the city's comprehensive plan). These range upwards to as much as 30% for new development in residential zones, with a high level of existing canopy.

The Asheville ordinance is triggered by any new development needing a permit, or any new proposed open uses (such as a parking lot or sports field). The standards also apply to redevelopment or building expansions over 1500 square feet. To see what canopy percentage an applicant has to reach, one needs to consult a table in the code.

Developers can take one of several approaches to comply with the tree code, according to Chris Collins, Planning and Development Division Manager for the Asheville Planning Department.[2] They can "preserve, plant or pay a fee."

The ability to pay a fee-in-lieu of is controversial to some, but a recognition of the difficulties of reaching these standards on denser developed parts of the city. The city charges range from $4.54 to $31.44 per square foot of tree canopy needed to meet the standard. Collins estimates the city has already collected "hundreds of thousands of dollars," and is set to collect close to half a million alone for one large project currently in the development pipeline. "We can take that [money] and we can acquire land, we can plant trees, give grants to private citizens to plant trees," says Collins.

A useful aspect of the Asheville approach is an explicit effort to create an incentive in the standards to preserve existing trees over planting new ones. Specifically, there is a sliding scale in the ordinance that lowers the overall canopy percentage where an applicant preserves existing trees. For instance, in Suburban areas with high existing canopy, if a developer wants to reach the required standard only by planting new trees, the required canopy is 30% of the site, but it drops to 15% if the applicant preserves existing trees on location. The code allows for combinations of new planting and preserving existing trees, but again gives more incentive to reach the goal by saving trees. Chris Collins tells me he believes these incentives are working. "So far what we're seeing is the incentives-based [system] is working so far. We're seeing the skew towards preserving trees ..."

For newly planted trees, one of the problems in Asheville has been that many end up later dying and that the city does not have a good system or way to inspect and monitor the health of these trees after they are planted. Often the planting job is farmed out to a subcontractor, who later is hard to find should the trees die. The failure to stipulate minimum soil volume for new trees is another part of this problem, notes Vaidila Satvika, also with the city's planning department, dooming these trees. These are fairly consistent issues for many cities attempting to enforce such regulations.

Tree protection standards are often connected to specific geographical areas, for instance a floodplain or ecologically sensitive area, and that makes some sense. Portland's code imposes specific tree protection requirements for development within its river overlay zone, for example.

Many cities also typically require measures to protect trees during the construction or development process, for example ensuring that heavy equipment or the stacking of building materials does not occur around the base of trees damaging its root zone.

How stringent tree protection standards should be is an open question. There are understandable concerns about whether and in what circumstances such restrictions curtail the building of new housing and the affordability of the overall housing stock in a community. It is the position of this author that it will usually be possible to protect trees and to further grow a city's tree canopy at the same time that new housing is accommodated.

That said, there is little doubt that conflicts will emerge, and difficult decisions and tradeoffs will need to be made. Cities need new housing, especially and primarily more affordable housing, that is true. But we must also keep in mind that healthy and livable neighborhoods will require abundant trees, increasingly so as cities become hotter. Privileging the construction of new housing to the detriment or loss of trees is short sighted and unacceptable. And in most cases there will be alternatives to this outcome.

Tampa, Florida, is one city that has a deserved reputation for protecting its trees, having adopted a tree protection code back in the 1970s. Fast forward to 2019, when the city embarked on a major re-write and update of that code, driven in part by growing frustration among developers about what was legally permissible under the code, a growing private property rights movement, and lax enforcement on the part of the city.

Things in Tampa came to a head when in 2018, at the behest of the Tampa Bay Builders Association, the city council voted favorably on a proposal that would have essentially allowed developers to cut down trees anywhere without restriction. Chelsea Johnson, who was president of her neighborhood association, and a lover of trees, was astounded when she heard of this proposal and sprang into action, eventually convincing the city's planning commission to vote against the clear-cutting proposal and founding the community group (cleverly named) "Tree Something, Say Something." Johnson and her group then spearheaded a unique year-long process of discussion and negotiation between the development community and the tree protection advocates, resulting in a new and innovative tree protection code.

Johnson describes herself as "the accidental tree lady," in the sense that while she has been a life-long lover of trees, she did not really think of the city's trees as being endangered.[3] But the council's action on the clear-cutting proposal was a wake-up call.

There are many pressures on trees in Tampa, as in many cities. In the established neighborhoods of the city many smaller homes were being replaced with much larger structures, and trees frequently got in the way. She is not unsympathetic to builders who find themselves with lots where a tree right in the middle would essentially make the lot unbuildable; "unless you designed a donut structure" she told me.

Johnson convened these negotiating sessions every Wednesday: in the morning she would host the developers, and in the afternoon the environment and tree lovers, who would all sit around the same dining room table. Johnson would carry proposals back and forth, until the two sides eventually agreed on a set of tree standards that would at once give more certainty to the development community, and significantly protect trees. It was a compromise, and one that neither side seemed completely happy about (which Johnson suggests might be the sign of a good compromise).

How Stringent Should Tree Codes Be?

A key issue of heated debate had to do with when a tree should be protected, and when a tree was located on a lot in such a way that its disruption was so great as to justify a developer cutting it down. "At what point are you supposed to accommodate the tree on your lot, that was there first, to the best of your ability, and at what point can you build a reasonably-sized home," said Johnson.

The essence of the compromise in the code is around the precise dimension of what the code calls a "tree removal zone," the zone defining the center of the parcel in which a tree cut be removed. Developers wanted a larger zone, conservationists a smaller zone, but both sides seemed to accept that such a defined space made sense. From the developer's perspective this would significantly and helpfully provide them with much more certainty when buying a lot about what was possible. As Johnson points out, the delineation of this zone (what the tree lovers on her side dubbed the "death zone") would also in turn delineate where trees should be protected.

"Ultimately," Johnson told me, "Compromising will save more trees." What resulted was a code that narrowed this tree removal zone: it takes the regular zoning setbacks and adds five feet in front, five feet on the sides and ten feet in the rear. The code offers substantial flexibility in diverging from the zoning setbacks, allowing a building or developer to more easily move the building site and or design the structure to avoid loss of a tree. Specifically the code allows for up to a 40% variance from the rear yard setback and 25% in the front yard, as well as a foot on either side of the lot.

There are other positive improvements, including that when new trees are planted to mitigate or compensate for trees cut down, these new trees can be planted on private property (not just in the public right of way). Also the required compensation is based not just on the trunk size of the tree cut down but rather on the size or width of the canopy of the tree, resulting in mitigation requirements that more accurately replace what is being lost. "We'll see more trees planted," Johnson says.

Not every tree lover would support these amendments, certainly, but Johnson feels on balance the result is fair to all parties and better for trees than the status quo. I ask why the need to modify the existing 1970s era code that has worked so well for so long to protect trees—she points to the massive political changes that have taken place and the growing support for private property rights in Florida, as well as an anemic, to put it mildly, enforcement system.

Even with strong code provisions, enforcement is often the Achilles heel. Johnson notes that there are only two staff responsible for code enforcement, not nearly enough, and a tree removal hotline that collects messages. "By the time tree enforcement got out there the tree would be gone," she told me. And much of the tree loss happens in more hidden ways, through a process she described as "chuck in a truck," where trees are cut and removed without permits by unlicensed and unethical contractors, flying under the radar of the community. "You drive by one day and the tree is there" she says, "and the next day it is gone." She notes that at least in Tampa the fines are small, maxing out at $15,000 (a small price to pay when many of the homes are valued at $1 million or greater).

Literally just a few weeks after Tampa had reached this historical compromise, the state's conservative politics became evident in the passage by the Florida legislature of a state preemption law. Passed also in 2019, the new state law allows any property owner to cut down a tree without need of a local permit, as long as a registered arborist signs off that the tree represents a "danger to persons or property."[4]

Other cities have sought to take similar steps to make saving larger, older trees easier again by allowing flexibility in the implementation of zoning and development codes. Seattle's approach has been to establish a clear category of legal protection for so-called "exceptional" trees: these are trees defined in the code as "rare or exceptional by virtue of its size, species, condition, cultural/historic importance, age and/or contribution as part of a grove of trees …"[5] There is a size threshold of thirty (30') feet in DBH, "or 75% of the largest documented diameter for a tree of that species in Seattle, whichever is less."[6]

On already developed parcels in the city it was forbidden under this code, to remove any exceptional trees, unless it is determined to be a "high-risk hazard." Like many local tree codes there is also a yearly limit on the number of smaller trees that a homeowner or property owner can cut down—under the Seattle code the limit has been three non-exceptional trees, or trees defined as trees 6 inches in diameter or larger (and again in the case of high-risk trees, these trees are not subject to the limit). Trees removed in Seattle, when allowed (exceptional and those 24 inches or greater in diameter), had to be replaced "with a tree that will provide the same canopy coverage at maturity."[7]

Defining when a dead or dying or diseased tree becomes hazardous, and justifies removal, is an issue dealt with in many local codes. Ideally a city's tree code should lay out a clear set of criteria or guidelines that are defensible: only when a tree is dead, or dying, or a threat to public safety, ought approval to remove be granted, at least if the tree is of considerable size.

Preserving Trees by Allowing Development Flexibility

At the heart of the Seattle approach has been the intent to provide strong protection for exceptional trees and a presumption that property owners and developers will be able to find ways to build around them. The city has explicitly allowed deviations from its zoning and other development codes, to make this possible. These deviations depend on zoning and density categories—for neighborhood residential zones, where there is a 35% limit on lot coverage—front yard and rear yard "departures" must be used to avoid tree removal. For low rise developments, proposals are required to go through a "streamlined design review," and must utilize a combination of "adjustments," principally adding height (to allow the building to take a more vertical shape, adding an additional floor for example) and /or reducing the number of parking spaces required. Only when these adjustments have been taken (explored) and the applicant still is unable to reach the maximum allowable floor area will a permit be issued to permit removal of an exceptional tree.

One concern about the Seattle approach expressed by tree advocates there, and really a universal problem, is that enforcement of the code, especially for single large trees in residential areas, relies on a kind of honor system. One has to know of the protections for exceptional trees and be willing then to take steps necessary to reconfigure the project to protect the trees. Oftentimes it seems that a homeowner or small-scale developer will not even know about the requirements.

In discussing the Seattle tree protection requirements with representatives from *The Last 6000 Campaign* it was their strong belief that rarely does the city require developers to go through the steps to see if by taking advantage of this extra flexibility in the development standards (e.g., add height to a building, relief from setbacks and parking) could save trees on site. What usually happens, they tell me, is that whatever the developer or landowner proposes is usually what is accepted. Partly this may reflect not only problems in the development review process but also limited staffing. Sandy

Shettler, of *The Last 6000 Campaign*, pointed out that while there are 460 employees of the Seattle Department of Construction and Inspections, there are only two arborists.[8]

In this sense the code seemed to depend heavily on developers to follow the standards, something difficult when there seems little culture or tradition or precedent for finding creative ways to work trees into a development or redevelopment project. It is possible, although leaders in *The Last 6000 Campaign* could come up with only one recent example of this. It is the positive case of Bryant Heights, a 2018 redevelopment project in the Bryant-Ravenna neighborhood of the city. A combination of single family, townhomes, and condos as well as some commercial space, and designed by Johnston Architects, the project preserves and celebrates large trees on the site. Specifically, the configuration of the development allowed the protection of twenty-five trees on site, "including a cluster of Douglas Fir trees that were incorporated into the overall design."[9]

> Circulation and pedestrian paths were incorporated into the design to enhance the existing pastoral landscape and take advantage of two large exceptional trees that became the major focal point of the internal courtyards 'central greenbelt.' The site-lines of the two exceptional trees along with the pedestrian circulation helped blend the 3-different building zones together to help create a unified community that compliments the wonderful Bryant neighborhood.

The trees then help to deepen the connection to community and landscape and their preservation and the prominence they are given in the site plan add an important dimension of nature to this project.

It is a fitting reference to the deeper history of the neighborhoods as a dense forest. As Sam Hart notes, writing about the project, when the town of Ravenna was annexed by Seattle in 1907 it "was still covered in old-growth forest." Bryant Heights in its incorporation of the trees is a small step in protecting and preserving the character of this community. Can we redevelop urban neighborhoods like this, densely and compactly, but also protect their ecology? Hart poses well the question many cities are facing: "How can we design projects that both expand and improve a neighborhood while still paying homage to the spirit of the community?" It is not so unusual, and I have seen in some of the most forward-looking urban neighborhoods, in cities like Freiburg, Germany, begin with protecting the older existing trees and designing urban life around them (Figure 3.1).

Cities can also regulate the companies, small and large, that are hired to cut down trees. To its credit Seattle city council adopted in 2022 an ordinance aimed at addressing a problem that some have called "predatory

FIGURE 3.1 A Children's Play Area in the Interior of the Vauban Neighborhood, Freiburg, Germany. Photo Credit: Tim Beatley.

cutting," or the largely unregulated small tree removal companies operating in the city. The ordinance now requires all companies providing tree care be registered with the city, to renew this registration yearly, and establishes a citywide "arborist registration requirement" as well. The city plans to publish and make public this registry. Companies operating in the city will need to have at least one employee (or one person on retainer) who has arborist credentials, as defined by the ordinance[10] (Figure 3.2).

At least in theory such a system would make it less likely that a tree company or tree service provider does not know the law or is unwilling to follow the law. It provides a mechanism for educating about the code's requirements and especially about the protections for trees defined as exceptional trees. And it makes it more likely that assessments of what constitutes a dead, diseased or hazardous tree, will be made professionally and accurately. One provision of the proposed ordinance that I find especially attractive is a requirement for a second independent assessment in the case of an exceptional tree where removal is being requested.

More specifically:

If the tree meets the City's definition of exceptional, the registered tree service provider shall engage another registered tree service provider to independently assess the tree and submit the application for its removal.

FIGURE 3.2   Many Trees Are Cut by Small, Unlicensed Contractors with Limited Tree Knowledge. Photo Credit: Tim Beatley.

The registered tree service provider that submits the application must be different from the registered tree service provider that will perform the removal or major pruning of the tree.

The presumption of the code that requires an "independent" second review of if and when a protected tree meets the criteria of hazardous, is a good idea I think, though there are obvious questions about how this would work, and whether (especially in smaller cities) companies would end up collaborating in ways that undermine the intent of the law (does one company rubber stamp another company's assessment, and vice versa). I find myself thinking about the so-called independent property assessments process in the US home mortgage system, where miraculously the value of homes are almost never assessed at below the amount of the loan being asked for. Every actor in the system implicitly knows what needs to be done for each to earn a profit. Nevertheless, these are procedural steps that offer some additional safeguards and certainly ought to be considered by other cities.

In speaking with the leaders of *The Last 6000 Campaign* they told me some alarming stories of unscrupulous tree cutters targeting homeowners with large trees, knocking on doors and essentially encouraging them to cut down these trees before they exceed the 30 inch "exceptional tree" standard, thus perhaps making it hard to remove them. One hopes this happens rarely, but I know from firsthand experience that small tree cutting firms frequently canvas neighborhoods and do often aggressively knock on doors too often mixing a potent message of fear and worry and promise of a discount price. A city's tree canopy suffers unfortunately from this dynamic, and from the undeniable fact that there is commercial value in cutting down, rather than preserving, trees.

In May 2023, as the writing of this book was coming to an end, Seattle city council adopted substantial revisions to its tree code, the first in over twenty years. The new code expands the number and size of trees covered by the law (with trees as small as 6 inches in diameter receiving some level of protection). The new code distinguishes importantly between tree protection outside of development review and requirements that would apply during development review. Most tree removal would be prohibited, say on a vacant lot or an already developed home site, but for new development proposals the requirements are different. Like the earlier code, developers would need to show that saving trees is not possible even when relief from setbacks and parking standards is taken into account.

Sandy Shettler and other tree advocates are unhappy with the new code because they feel it will essentially make it easier for developers to cut down existing trees. It is not clear that the new code is less stringent in this regard, but Shettler and others see a missed opportunity to strengthen the code to better protect existing trees in the development process. She expects developers will mostly cut them down and satisfy the (new) replacement requirements—planting small trees elsewhere to compensate or paying a modest fee in lieu of this. "Are we truly protecting trees," she asks in a handout she submitted to city council: "The new tree ordinance ... helps us plant more trees," she writes, "but misses the mark by encouraging development practices which remove trees from the places where people live: in our residential zones."[11] She'd like to see a code that resulted in more projects like Bryant Heights.

Holding the Line on Older Trees

And in many cities, there is considerable economic profit for developers and builders in cutting those trees down, often viewed (inexplicably to me) as an obstacle (rather than an asset). It is inconvenient to save trees, even protected trees, and easier to ignore the law and pay a modest fine. Or even a large fine that represents a modest cost of doing business in lucrative urban markets where new homes generate tidy profits.

Often it seems that even in cities with strong tree codes on the books, there is a pattern of property owners and developers ignoring the code, and simply paying whatever fine or fee is applied later. This has happened recently in Washington, DC.

Washington DC's approach dates to 2016, when Mayor Bowser signed the Tree Canopy Protection Act. Under this Act two categories of trees, defined by relative size, gain some level of protection. "Special trees" are those with circumferences between 44 inches and 99.9 inches, while the more stringent protection applies to Heritage Trees, or trees with circumferences of 100 inches or greater. Tree removal permits are required in both cases, though for smaller special trees where no hazard exists (that would justify cutting the trees) this essentially requires landowners to pay a replacement fee (at $35 per inch of circumference) or plant young replacement trees ("a quantity of saplings whose aggregated circumference equals or exceeds the circumference of … trees to be removed").

For Heritage Trees, a removal permit is to be approved only when "necessary to avoid imminent harm or danger to person or property," and requires an arborist to attest to this. It is a stronger standard and intended to prevent the cutting down of these older mother trees (Figure 3.3).

One option under the act is to relocate a Heritage Tree. This has happened infrequently, but a notable case, the first one under the law, is the relocation of several large trees from the site of the former headquarters of Fannie Mae. Being redeveloped as a mixed-use residential, office and commercial development, one of the large trees would sit now on the driveway of the new Wegmans grocery. So, in 2019 these trees were relocated, through a remarkable engineering feat, utilizing a platform of metal bars and inflatable tubes. The arm of a CAT excavator pulls the tree and its platform in the desired direction as it rolls over the long-inflated tubes. Reporters for WAMU captured the moving day on videotape, speeding it up the tape to better convey the methods and steps in this impressive move.[12]

Moving a tree that is estimated to weigh around 600,000 lbs is a costly endeavor, in this case each of the trees moved was estimated to cost around $200,000. Yet, the developer sees this as a way to protect the sense of place and the natural heritage of the site. "[W]e're trying to create as real a place as we can," he notes, "as authentic a place as we can"[13] (Figure 3.3).

The enforcement of DC's tree provisions has been troubled (and troubling) to say the least. One pattern has been for developers and property owners to essentially ignore the act, cutting down Heritage trees and then simply paying fines. The fines are hefty ($300 per inch of circumference under the Act) but appear to be treated as a cost of doing business and economically justifiable given the high profits to be made in land development in the District of Columbia.

FIGURE 3.3 A Heritage Tree in Washington's Glover Park Neighborhood. Photo Credit: Anneke Bastiaan.

While this recent legislation has strengthened the act, there remain challenges. What to do when situations arise that seem to pit tree protection against affordable housing? A notable recent example can be seen in the case of a proposed 200-unit development apartment complex, part of a larger development called Parkside. The developer, arguing that the city needs more housing, has sought an exemption from the tree protection law arguing that the city needs more housing. One councilmember (who represents the site where the project would take place) has been receptive and has introduced legislation to provide the exemption, while others, notably (now retired) Councilmember Mary Cheh, have been clear in their position that that tree must be protected. To complicate matters, representatives for the developer also argue that the entire project received development approval some years ago, prior to the enactment of the Canopy Act.

There may be an argument from fairness for some sort of remedy in this case, but providing exemptions from the act is a slippery slope and a bad idea. Councilmember Cheh has been clear about her concerns: "I was not inclined to consider an exemption ... I know the exemption business: Once you have one, then you have two."[14] Five trees that old and that large ought to be greeted with wonder and seen as an opportunity. And it is also surprising that the shade and other benefits these trees bring to the neighborhood are not immediately seen as of equal value, or necessary complements, to the housing.

Parkside is indeed a rather large development, and the renderings indicate lots of (albeit smaller) trees and an overall natureful design (including green roofs on several of the larger buildings). There would, on the face of it, seem to be many ways to shift around the density in this development; to add a floor or expand one building's footprint to make up for whatever lost developable space results from saving the trees, though the developer would likely object to such changes so late in the game.

Balancing Trees and Property Rights

In many cities the tree protection codes seem mostly about allowing developers and property owners to remove trees and in exchange to require replacement trees or the paying of a small mitigation fee to support new trees. Many of these codes are not very strong, then, and seem almost designed to facilitate tree removal rather than tree protection. That is too bad, though the political will in many cities seems to be shifting in the direction of stronger codes. The strengthening of DC's law is a good sign, and there is much interest in other cities for stronger measures to protect trees on private lots.

How to save a large tree, or grove of trees, when by doing so the building lot becomes unbuildable, or able to accommodate a significantly smaller number of units, remains a real challenge to urban tree conservation. Providing greater flexibility in design and siting as the Tampa and Seattle examples demonstrate is one part of the answer. But what happens when, as in the Parkside case earlier, protecting the trees means development of the site is difficult or largely impossible?

Tree conservation advocates might reply that there is no inherent right to develop if it means killing an ancient tree or trees, providing significant public benefits. There are many vacant sites in a city usually where few if any trees are to be found. Or perhaps you build what you are able to build on site, given the need to protect the trees: perhaps it is a smaller home or building, or a use ancillary to an adjacent lot. Much of the pressure on trees in many cities arises from the phenomenon of "tear-downs" where smaller homes are replaced with larger homes. Perhaps in these circumstances a property owner or home builder ought not to be entitled to any larger structure than what was there before if replacing with a new larger structure would necessitate the removal of a significant tree? And in every case, it makes sense to ensure that all design and siting options have been considered before a tree is allowed to be removed.

There are as well some important tools that cities are experimenting with, adaptations of tools that have actually been used for many years for other conservation goals—notably to protect historic buildings and prime farmland. There are perhaps other ways to address this conflict, for instance, purchasing the parcel and compensating the property owners for their economic loss.

Some cities have utilized the tool of Transfer of Development Rights (or TDR) to address these kinds of circumstances, though it has not been used to address tree protection. This tool has a long and storied past and has been especially useful in protecting farmland and historic buildings. The basic idea is quite simple: protect an existing parcel of farmland or an historic structure by allowing its owner to transfer the unused density to another location where it is desired. TDR programs vary in their specifics; in some places the locality serves as a broker, buying rights from one landowner and selling them to a developer for use somewhere else in the city. The idea requires a city to designate both sending zones (those areas of important farmland or environmental conservation areas where development is to be discouraged or prohibited) and "receiving zones," or those parts of the city where the acquired rights can be used, usually to increase permissible density.

Portland is one such place, a city with a long history of efforts to protect urban trees. It has recently faced special pressures in the eastern part of the city, where lower density development patterns mean that in many places there are still remaining relatively large groves of native trees, such as Douglas Fir. To make it easier to save such trees, the city is one of the first to have adopted a new Tree Preservation Transfer of Development Rights provision (or just *Tree Preservation TDR*, for short). The idea is that by setting aside a large tree, or even a grove of trees, you are then able to transfer the unused density from that site to another parcel in the city where it can be used to increase allowable density. In this way, there is a mechanism for at least partially compensating the property owner for saving the tree or trees on her property.

The extent of the transferable density will depend on the diameter of the tree or trees, as well as the development zone. The larger the tree the greater the transferable floor area. Once the transfer occurs there must be covenants attached to both the sending and receiving property deeds. In the case of the sending parcel the trees must be protected for at least fifty years.

In Portland, the provisions are so new they have yet to actually be used. One issue is that there is currently no mechanism for "matchmaking" to occur—that is linking a property owner interested in saving a tree and transferring (or selling) development rights with a developer or property owner somewhere else in the city interested in using these rights. In some places the city serves as a more active broker, and may even buy and sell the rights itself.

It might work well in the case of DC's Parkside development, allowing the developer to shift the density to another place in the city, perhaps another parcel in the same neighborhood, and in this way mitigate the economic impact while preserving the trees.

The use of TDR for tree protection is unusual, though Portland, has amended its zoning (in 2020) to make this possible. Under its "Tree Preservation Transfer of Development Rights" provisions it is now possible to transfer density in the form of unused FAR (Floor Area Ratio, a way that many cities regulate density) in exchange for protecting, for at least fifty years, trees and groves of trees in the city.

The amount of the permissible FAR that can be transferred depends on the number of trees to be protected and their size as well as the zone in which the trees sit. This unused density can be transferred to any multi-unit zoning district in the city, with the exception of locations in the Central City. The transfer (and resulting conservation of the trees) is implemented through deed covenants placed on both the sending and receiving parcels.

The Promise of Tree TDR?

Can this TDR option help to address those situations where preserving older existing trees make it difficult to develop, or fully develop, a site? So far, there have been no takers yet and no use of this TDR program to preserve trees in Portland, though again the provisions are very new. "But without City efforts to educate or raise awareness about this, the opportunity will likely sit unused," says Kyna Rubin of Trees for Life Oregon, a local tree advocacy group.[15]

"The biggest challenge for Portland," Rubin argues, "is connecting sellers and buyers of development rights—matching property owners who want to preserve their large trees by selling development rights with builders who'd like to add a floor to a structure on an arterial street. How to find a local champion who will guide parties through the process? If the City's not going to take the lead, who will?"[16]

I spoke with Bill Cunningham, with the Portland Bureau of Planning and Sustainability, and who runs the TDR program. He agrees with the critique that the city needs to do a better job getting the word out about the option. "We need to get people aware of the provisions," he said.[17]

Cunningham mentions that Portland has been inspired and guided by King County, Washington, which has operated a relatively large TDR program for many years, aimed especially at protecting farmland and forestland, mostly outside the city. King County does in fact take a more active role, buying and selling development rights.

Cunningham and I talked specifically about the eastern part of the city where because of its unique development history, tends to be the location of larger groves of Douglas fir, and areas especially suited for development transfer. For Cunningham, the use of TDR to preserve historic buildings is a fitting background and believes there is an increasing sense that trees are essential to preserving the history of these urban landscapes.

Part of the answer must surely be to think carefully about the trees we have and to take steps early in the development and design process to minimize the need to cut down trees in the first place. Greg Levine, co-executive director of the nonprofit Trees Atlanta, believes this early approach of early designing with trees in mind is critical. He described a case of a library where the plans called for the loss of five large shade trees, something many in the community objected to; this might have been avoided had the building been designed from the start with the trees in mind.

Levine notes that ultimately for codes and tree protection standards to be enforced requires strong commitment from the top, and he agrees with others interviewed in this book that a new attitude about trees is needed—one which resists these tradeoffs that seem always to result in the loss of trees. He'd like to see more "lines in the sand" when it comes to protecting the trees we have. Unfortunately, there are few cities that seem willing to draw such lines in the sand.

Lines in the sand are hard politically because they are often resisted vehemently by developers and property owners, and such codes if enforced certainly do limit what is possible on a site or parcel. Yet, as Levine suggests, tree protection should increasingly be understood essentially as the parameters in which developers and landowners operate. "Developers need to work within these parameters, and there's still a way to make money and still a way for people to move to the city," he told me.

One thing this shows is that the job of tree protection simply does not stop when the code or ordinance is adopted. Effective enforcement and implementation will require much more. It will require adequate staffing levels and budgetary commitments on the part of cities. And it will require education and active efforts to reach out to constituents, neighborhoods and neighborhood organizations, and particular community interests that you want and need to have on your side. This includes developers and real estate professionals, especially. Chelsea Johnson in Tampa set a good example in that regard, recognizing the need to engage the development and real estate community in discussions about the tree code. If there is broad agreement and personal commitments the code offers a much better chance of succeeding in protecting and hopefully expanding a city's canopy.

Conclusions

Cities have adopted tree protection codes and standards that vary greatly in their details and stringency. Every city should have a minimum set of tree protection standards, and to my thinking they ought to be as strong as they can be, given how important trees are for the health and wellbeing of residents. Enforcement staff and budgetary commitments to enforcement will also be essential.

I have lived for multiple decades in a city without tree protection standards to speak of and have been occasionally shocked by the callous and careless ways in which trees have been damaged or cut down here. Several months ago we were shocked to see on our daily walk through the neighborhood a majestic and quite large pin oak tree being cut down. It was at least to us a very healthy tree, one that we enjoyed seeing and watching over time, and its large canopy was home to many birds. We will never know how or why the decision was made by the homeowner to cut down this tree, but I have wanted to better understand it. This large tree would have required a permit under most of the tree codes reviewed here. While the tree might still have been lost, at least there would have been a pause and some form of deliberation about its fate, and at least some recognition of the public dimensions and impact of that homeowner's decision. Tree codes can at a minimum provide such a pause. Even when loosely enforced they send the signal that trees are in a special category of things deserving of care and protection, and only when there are serious and important reasons ought we tolerate their being damaged or cut down. Codes are both a reflection of what a city or community values as well as tangible and visible expressions of these values; they are moral signals or ethical cues that remind us of what is valuable and important in our lives in and in our communities.

We regulate many things for important reasons, and as a result we place reasonable restrictions on personal freedoms. Submitting to a tree code and to limitations on how and when you can cut down a mature tree seems little to ask of a homeowner. Where the city mandates the protection of trees to achieve a larger public good or advance public health it is also not unreasonable to take steps to lessen the personal burden. Tools like Tree Preservation TDR can be used and, in some cases, public purchase of a parcel when tree protection makes it difficult to productively use one's parcel.

But codes like these establish a city's minimum rules and expectations. They are the outcome or expression of a city's values, but they also send signals about what is valued and important in a city, and what our collective responsibilities are to protect and sustain these trees as part of common ecological assets and collective heritage.

Cities will also need a set of reinforcing policies and incentives. More about this will follow later, but a tree code will meet with less resistance if there are financial incentives that go along with regulations. For instance, the idea of some form of treebates, where residents receive grants or perhaps reductions in property tax bills that reflects the positive value created and provided for the larger public by saving trees.

Some of the city codes allow for the designation of protected "groves." Los Angeles' code allows for this, for instance, and while enforcement has equally suffered, there may be an advantage to designating larger groupings of trees in cities. As another example Washington, DC has now begun designating forest "patches" in the city and while there is little legal protection that goes along with this designation, the city will put up signage at least (as it has done in the case of the Langdon Park Forest patch) and this kind of step will likely be helpful in giving attention and political status at least to trees in cities. Giving trees names is one idea I explore later, but even better if a "forest" can be so named and identified. In this case it may be easier to defend the trees.

Notes

1 See "Heritage Trees," San Francisco Department of Environment, found here: https://sfenvironment.org/landmark-trees, accessed April 20, 2023.
2 Interview with Chris Collins and Vaidila Satvika, Asheville Planning Department, 2022.
3 Interview with Chelsea Johnson, Tree Something, Say Something, 2019.
4 Charlie Frago, "DeSantis Signs Bill Weakening Tampa's Tree Ordinance," *Tampa Bay Times*, June 19, 2019, found here: https://www.tampabay.com/florida-politics/buzz/2019/06/27/desantis-signs-bill-weakening-tampas-tree-ordinance/, accessed June 6, 2023.

5 See Director's Rule 16–2008, "Designation of Exceptional Trees," found here: https://www.seattle.gov/documents/Departments/UrbanForestryCommission/2018/2018docs/dr2008_16x.pdf, accessed July 28, 2022.

6 Ibid.

7 See: "Tree Protection Regulations in Seattle," found here: https://www.seattle.gov/documents/Departments/SDOT/Trees/cam242.pdf, accessed July 28, 2022.

8 Interview with Barbara Bernard, Jim Davis, Jessica Dixon, and Sandy Shettler, The Last 6000, August 8, 2022.

9 Fazio Associates, "Bryant Heights," found here: https://www.fazioassociates.com/bryant-heights

10 The code states that a tree service provider must have "at least one employee or a person on retainer who is a currently credentialed International Society of Arboriculture (ISA) certified arborist trained and knowledgeable to conduct work in compliance with American National Standards Institute (ANSI) Standard A-300 or its successor standard."

11 Sandy Shettler, "Are We Truly Protecting Our Trees with This Legislation?" undated.

12 Jacob Fenston, "How (and Why) Do You Move a 600,000 Pound Tree," *WAMU*, January 23, 2019, found here: https://wamu.org/story/19/01/23/how-and-why-do-you-move-a-600000-pound-tree/, accessed June 2, 2022.

13 Ibid.

14 Jacob Fenston, "Developer Seeks Exemption to Heritage Tree Protections, Citing DC's Housing Crunch," *DCist*, May 16, 2022, found here: https://dcist.com/story/22/05/16/developer-exemption-heritage-tree-dc-housing-crunch/, accessed May 19, 2022.

15 Kyna Rubin, "Invisible Incentive to Preserve Large Trees," *Trees for Life Oregon*, found here: https://www.treesforlifeoregon.org/invisible-incentive-to-preserve-large-trees, accessed May 18, 2022.

16 Ibid.

17 Interview with Bill Cunningham, May 26, 2022.

4

MANAGING THE URBAN FOREST

Managing the public (and private) forest in cities raises numerous challenges. Keeping newly planted young trees alive is a special challenge and growing the urban forest and making meaningful progress toward a city's ambitious canopy targets. Cities face pressing questions as well about which species, or mix of species, to select as they expand their urban forests. The explosion in new research about biology and trees and forests is creating both confusion and wonderful new opportunities. While the planting of non-native species of trees is still occurring, there has been a growing awareness of and argument for planting more native species. Sometimes this argument is rooted in the circumstances of a changing climate, in other cases in what we increasingly know about the ways in which native trees support wildlife and biodiversity.

The 10–20–30 rule in forestry, mentioned in Chapter 1, reflects this need to plant a diverse urban forest (e.g., plant no more than 10% in a single species). Too often cities plant many of the same species, often non-natives, perhaps out of fashion, familiarity, or convenience. Tree-killing diseases, such as Dutch elm disease, or pests such as emerald ash borer in North America, show compellingly the vulnerability of an urban forest comprising a few species however appreciated or beloved by residents or designers. There are increasingly a range of different considerations cities must take into account in choosing what and where and how to plant trees and forests.

DOI: 10.4324/9781003377344-4

What Should We Plant and Grow?

There is no question that the existing trees and forests in cities will be stressed by climate change. One recent study of more than 3000 tree species in 164 cities around the world, concluded that many of them will exceed their natural "safety limits" in terms of the extent of heat and drought they can tolerate when future climate projections are taken into account.[1] The hardiness zones of North America have already been shifting and will continue to do so.[2] Many cities are already beginning to plant different species based on changing climate, for instance North American cities choosing to plant tree species that are more common further south. And there will be significant and serious management challenges, the need for example to ensure adequate watering for many species in the face of increasing periods of drought (more on that later). And these climate stressors will affect trees in different parts of the city: largely standalone trees along an urban street may experience greater levels of heat, than say trees located in parks or larger patches of forests in a city. The planned location of a tree in the city will also influence the choice of species.

The new reality of wildfires in the era of climate change suggests additional conditions in the choice of tree species (as well as in the management of urban and semi-urban forests). Species with thicker layers of bark, for example, will tend to be more fire-resistant.[3] In the Western US, native species like coast live oak or ponderosa pine are more resistant to fire, while non-native such as eucalyptus are highly susceptible to burning.[4] Resistance to fire is yet another reason to protect the larger, older trees, as they tend to have thicker bark and deeper roots, and greater distance to the tree crowns from flammable ground materials.

Trees can also be selected based on how fast they grow and how quickly they sequester carbon. A recent Swedish study calculates the number of years several different species of tree must grow before they become carbon-positive: that is, when they sequester more carbon than was emitted to grow, transport, plant and maintain them (e.g., carbon emitted by trucks, water pumps).[5]

Planting trees that provide habitat for birds and other animals will also be a priority, University of Delaware entomologist Doug Tallamy has emerged as the most passionate voice for planting native trees. His research makes a strong case for the coevolution of native trees and many other species, notably birds, that depend on those trees. Native trees, like white oaks, provide an abundance of food for birds, says Tallamy in his book *Nature's Best Hope*.[6] Nesting birds mostly depend on caterpillars to feed and raise young, and native trees have evolved to serve as critical hosts to such lepidoptera species. The task of raising and fledgling a nest of songbirds is remarkably demanding, with Tallamy noting that thousands of individual caterpillars are required (by one study, up to 9000 caterpillars are needed to raise a nest of Carolina chickadees)[7]

Even within native plants and trees there is great variation, and Tallamy argues that at least some of the trees planted need to be what he calls "Keystone Plants": the approximately 5% of native plants (and trees) that produce some 75% of the caterpillars. "Without keystone plants, the food web all but falls apart."[8]

The number one Keystone Tree, and Tallamy's personal favorite, is the white oak (*Quercus alba*). He estimates that this species provides habitat to support a large number of birds. It is a host tree for a remarkable 454 species of caterpillars; these trees become essential in providing the food for raising young birds during nesting season.

It is not only the caterpillars but also the immense amount of food provided by the acorns it produces. In another of Tallamy's books, *The Nature of Oaks*, a passionate ode to the beauty and value of these special trees, he describes the symbiotic relationship between oak trees and blue jays.[9] The jays eat the acorns (and Tallamy tells us they have even evolved a beak designed to easily penetrate the husk of an acorn, as well an extra-large esophagus, called a gular pouch, for collecting and carrying the acorns), and in return help to propagate and disseminate the oaks. It is what Tallamy calls an "ancient mutualism." Tallamy estimates that a single blue jay will collect and bury as many as 4500 acorns in a single season, many he/she will forget to retrieve, something like three-quarters of the planted acorns he estimates. That leads to Tallamy's astounding estimate that each blue jay may plant as many as 3360 acorns that could later grow up to be mature oak trees. British writer Robert Macfarlane has noted recently just how many different species rely on oak trees. As he recently tweeted:"The English Oak (*Quercus robur*) supports an astonishing 2300 species, of which 326 depend on it for survival. 716 lichens, 108 fungi, 1178 invertebrates, owls, bats, wood warblers, butterflies... I hold in my hand not a single tree, but a community-to-be, a world-in-waiting." An oak acorn is, indeed, as he so eloquently writes, holds the potential to support, when it grows and becomes a mature tree, a remarkable abundance of life. Birds yes, but many other species comprise a true community of life.

Tallamy recommends that we profoundly rethink the yards and spaces around our homes. This is a wonderful and accessible first step to bringing native trees into our lives. In *Nature's Best Hope* he introduces the concept of the Homegrown National Park, noting how much space we have devoted to this mostly biologically sterile form of turfgrass lawn. If we each committed to converting just half our lawns into native plants and trees, Tallamy says, that would protect and create some 20 million acres of habitat and would amount to the nation's "largest park system."[10]

Every home needs multiple trees, and those trees can and should be integrated into the lives of the human occupants of the houses and sites in which they are planted. With native trees will come the chance to see and interact with many other wild animals. Many birds, yes, but also squirrels and chipmunks, raccoons, and other mammals that depend on trees, as do many smaller creatures such as arthropods. As Tallamy says, "Close interaction with the wild animals in your yard can bring you the same emotional benefits that are gained from living with cats and dogs."[11]

Allowing Trees in Cities to Grow Old

There is also a growing awareness of the need to protect larger, older trees in cities, which evidence suggests will be disproportionately more valuable in terms of sequestering carbon and providing important ecological services.

And there is now some evidence that the health benefits for urban residents will be disproportionately greater in the case of larger trees and trees with larger crowns. Precisely why this is so is unclear but it is indisputably the case that larger crowns will provide more shading as well as stormwater retention and habitat values compared with smaller trees.

A recent study (mentioned earlier) of residents of Brussels found an inverse relationship between medication sales for treatment of mood disorders and cardiovascular disease were inversely related to trees, but there was an especially strong relationship to larger trunk size and crown.[12] Another recent study showed the relationship between tree planting (over a thirty-year period) and non-accidental mortality, a relationship that holds over time as trees get larger and older. The authors of this study conclude that "our finding that larger trees are associated with greater reductions in mortality is consistent with several mechanisms."[13]

> For example, as a tree grows, its leaf area increases, which also increases the ability of the tree to absorb air pollution, moderate temperatures, and dampen noise. In addition, across cultures, larger trees are aesthetically more appealing, so larger trees may be more psychologically restorative, and they be more effective at promoting social cohesion.[14]

Suzanne Simard, a forest ecologist at the University of British Columbia, has dispelled this mythology, discovered that trees share nutrients and carbon through the elaborate underground mycorrhizal networks that exist and that connect trees in essentially a community.[15] While the implications for urban forestry have yet to be fully considered, they are great. Planting trees in closer proximity to each other and allowing (rather than preventing) these fungal and root interconnections to take place, may be essential to ensure that urban trees withstand and survive the stresses they will face.

How we manage existing urban forests also raises questions about how aggressively we cut or prune trees and what level of risk we are individually or collectively willing to live with. New perspectives on urban forests suggest we need to maintain more standing dead stock—these trees provide essential habitat for woodpeckers and other birds that rely on nesting cavities. We are often too quick in cutting down trees and extracting the biomass from the forest site even when for safety reasons some action might be justified.

If a tree represents a danger to a homeowner or to the public, say because it is sick or damaged, cutting down that tree might be justified. But there are different approaches that could be taken. Recently in my neighborhood a large, old red oak tree was hit by lightning. The consensus of the tree experts was that the tree would not survive (and a gaping split could be seen from top to nearly the base of the tree). The tree was cut into segments with the help of a crane and then entirely cut down. I wonder in such cases whether topping the tree and leaving much of the shank or base of the tree could be a better option. If the tree were to fall in this case it would not endanger either house on either side, and it could stand for many years providing essential habitat and continuing to store carbon. But for some reason, perhaps because of the look of a dead tree, this option is rarely chosen.

Leaving Room for Trees to Die

Cities should actively work to make room for dead and dying trees. The ecological value of a tree extends many years beyond its death. About a decade ago I recall interviewing a Stockholm researcher about his work on the importance of nature close to cities. We agreed to meet at Stockholm's Ecopark, an urban national park, that provides residents there with a dose of nature, including older trees. As we walked, he explained the Swedish philosophy behind the life and death of trees such as oaks. These ancient trees, it is thought, live for 500 years and then die for 500 years, he told me. In the "dying" there is much life that is nurtured and supported. Yet we eschew the look of decomposition, of trees and vegetation, even though this is as natural and essential a life stage as the growth of the tree.

There are several standing dead trees that I encounter on daily walks and I am always impressed with the sounds and activity emanating from them—woodpeckers seeking food, important nest cavities for a variety of birds, and high perching spots for the occasional Cooper's Hawk. And if we want to support birds, these trees are important habitat for many of the things birds eat—they serve as important overwintering sites for many caterpillar species (such as the giant leopard moth that I frequently see in my neighborhood).

Cities are beginning to understand the value of dead and dying trees and modifying their park management practices and policies to accommodate this. Less often do we see this as a priority on private property, perhaps out of a fear that that standing dead tree represents a hazard to property or health, a legitimate concern but often because it seems to many to be untidy or unkempt.

One prominent example can be seen in the Queen Elizabeth Oak, located in Greenwich Park in London (Figure 4.1). Several years ago, I sought out this ancient tree, planted in the 11th century and touching so much of England's history, learning that it could be seen here. Finding it was a bit anticlimactic. There is not much left of the oak. Even though it likely died sometime in the 19th, it remained standing until a storm overturned it in 1991. Unusually it has been allowed to sit in a horizontal condition, gradually decaying over time. There is now a sign that tells visitors about the oak. Today it is "covered in a wonderful variety of bugs and fungus," declares one history website.[16]

Another groundbreaking insight from Simard's work is the importance of what she calls "mother trees," larger, older trees that serve to come to the assistance of younger trees. These older trees are essential to helping young trees survive stresses like drought or disease. Their outsize importance further supports the need to protect them and to shift toward planting strategies that emphasize these older trees and place them at the center.

These larger trees we know now will store vastly greater amounts of carbon, compared with the earlier emphasis on the planting of new trees (and seedlings that often do not even survive into mature trees). Preserving these older, larger trees will accomplish other goals in cities, as we have seen. They are sometimes called "witness trees" because they existed in a place and time when momentous events unfolded. We can imagine and visualize these historic events more vividly when we witness trees around us who were literally present at the time the events occurred.

FIGURE 4.1   What Remains of the Queen Elizabeth Oak in London. Photo Credit: Tim Beatley.

There is new interest in managing urban forests in ways that enhance overall biodiversity and ensure a greater sense of wildness as well. There has been growing consensus about the need to focus more on larger species of trees—not only protecting these existing trees but also planting species that will eventually grow larger canopy and provide more shade. Portland, Oregon, planner Bill Cunningham mentions how that city is changing their planting standards to favor such larger species. Allowing a large Douglas Fir to be cut down and replaced with an ornamental cherry or smaller maple tree is not, he believes, a very good trade. "You're not replicating some of what is lost," he told me.[17]

Often in our attempt at keeping our cities and urban neighborhoods tidy we are quick to cut down trees that are dead or dying, leaving behind little of what was there before. Partly this is motivated by our prevailing aesthetics and the understandable desire not to be perceived as neglecting our homes and yards, or in the case of parks and public spaces, the responsibility of governments to ensure adequate upkeep of safe and clean public spaces.

Leaving space for trees to die and decompose helps biodiversity in many ways, of course.

Leaving dead and dying trees in the urban landscape is an important step adding essential habitat for woodpeckers and other cavity feeders. Yet, such a tree may strike neighbors as an eyesore and inconsistent with the pattern of tidy lawns and manicured spaces around the base of trees. Homeowner attitudes can change of course, and so also policies and covenants. Reston, Virginia's, explicitly allows homeowners to maintain snag trees. And increasingly even larger institutional projects in cities are being designed in ways that not only allow dead and dying wood but also celebrate it. A recent plan for the redesign of a section of the University of Oregon campus in Eugene, for example, includes an extensive "stumpery," making dead logs placed on the ground, also sometimes called "nurse logs," a visible and prominent part of the landscape there (they even become places to sit).

There are many threats to trees, of course, and many ways in which the life of a tree can be cut short. How aggressively to combat disease and threat is another open question, and how much cost to incur to ensure that our mature and older trees are allowed to grow older still, is another open question.

Many older trees are often outfitted with lightning rods, as lightning strikes do represent a significant threat. My prominent white oak, near to my academic office, has a lightning rod but surprisingly few of the older trees on my campus have been fitted with them. Why, I wondered. In asking our facilities management division, in charge of trees and landscape, I received an interesting and unexpected answer: the tendency for students and others to attempt to climb the wires. Perhaps it is a worry about legal liability, or about the safety of students, but it strikes me as mostly unlikely, and an insufficient reason not to invest in the installation of something that could save the life of ancient trees and the expansive biodiversity and nature in turn supported by these large trees.

Pests and disease are a serious issue, and climate change will exacerbate these problems. In many cities the urban forest is not very diverse, making it vulnerable. The emerald ash borer, for example, continues to devastate ash trees throughout North America. Yet there are treatments for these trees and at my own university they are being applied successfully. It involved drilling small holes at the base of the ashes, and injecting a pesticide that kills the borer. For the trees on my campus this is a treatment administered every two years and has been highly effective at saving these ash trees.

Especially in cities, there is every reason to be aggressive and to take all of the steps available to ensure that these precious larger trees survive. The many benefits provided especially by these larger trees will more than outweigh the effort and expense to control these pests and diseases.

Where and How to Plant

Cities will need to continue to diversify the species they plant. And impacts of climate change are causing many cities to rethink the species mix, recognizing that future urban trees will likely face more extreme temperatures, more extreme storms and weather events, as well as periods of stressful drought. Trees in many cities will need to be even more resilient that they already must be in cities.

A key question for any city will be where to plant trees and forests and how to practically expand and extend the canopy. There are many dimensions to this challenge and many interesting and creative options a city can pursue, ideally more than one at a time.

This is no academic question as many cities are setting ambitious tree planting targets for the future. Wellington, NZ, set the goal several years ago of planting two million new trees, and several American cities that have already completed million tree campaigns, are considering additional million tree goals.

In the limited spaces of a dense city, creativity may be required to find adequate spaces for these new trees. Options should consider tree planting integrated into rooftops, terraces and building designs, and repurposing parking and roadway space in cities to trees and forests (ideas taken up more fully in other chapters). Cities have many smaller, larger ignored and leftover spaces between and around buildings that could be places for forests and trees. One option for cities to consider is to look for places to plant very small forests. Japanese botanist Akira Miyawaki has made famous his methodology for planting very small forests that grow very quickly. The technique involves the layering of many native species, and planting them densely and in close proximity, each competing with others, and in the process growing remarkably fast (more detail about this option later).

Reaching desired canopy goals will require cities to look to other creative planting locales and strategies. As already discussed, more trees will need to be planted as part of building designs, on terraces and balconies, or even in the form of rooftop forests.

The Insurgent Urban Forest

Simard's work has helped to show how planting trees in neat rows, tree roots and trunks carefully separated and distanced, may not make much sense ecologically and in terms of the long-term health of the forest. Trees where possible, ought to be planted in groupings and groves, recognizing the social connections and ecological resilience such connections provide.

For many of us who live in cities the straight lines of street trees especially seem unduly rigid and linear and not especially natureful. Might we imagine more organic dimensions to where trees and forests in cities are planted or allowed to grow? The benefits and values of wildness that we cherish especially in urban environments will be difficult to achieve through a conventional street-trees approach.

There is also an ongoing and important debate about how tidy or controlled trees and forests are within cities. They are not usually permitted to grow, at least in already developed dense parts of cities, beyond the relatively straight and confined lines of streets and sidewalks. There has been recently a call in the urban planning community for a more "transgressive" form of urban forest—wilder, less tidy, less confined to the straight and narrow lines that street trees usually follow. Laurian, Sternberg and Voigt da Mata, in a provocative and thoughtful article in the *Journal of the American Planning Association (JAPA)* are critical of the typical urban forestry that gives "an impression of safety, bucolic and domesticated nature."[18] Another kind of aesthetics is required, they believe: one that "more truthfully depicts nature's struggle in the face of overwhelming assault."

We should seek to expose urban residents to trees and forests that exhibit an "unkempt" or overgrown wildness, they believe, and words like "escape," "invade," and "displace" are better ways to describe the new kind of trees and forests we could find in cities. Instead of neat and tidy rows of street trees we could imagine "forest shards" that jut into streets, forests that partially take over streets and parking lots, that might continue to grow and extend vertically up the sides of structures. In short, they imagine "insurgent trees" that "escape lawns and rights-of-way to march into civic spaces."

The authors continue in their description of these transgressive urban forests:

> *Insurgent trees* escape lawns and rights-of-way to march into civic spaces. They invade roadways (annoying and calming traffic), trample onto sidewalks (pedestrians must circumnavigate around them), penetrate parking lots (displacing parking bays), and obscure lines of sight. In some places, the street, plaza, and superblock are gashed with strips of soil, succumbing to arboreal invasion. In others, trees crack the concrete and buckle the asphalt. Vines ascend walls and gain footholds on green roofs. The newly forested vistas sway with winds and thicken and thin out with the seasons. In these ways, urban forestry presents a liberated nature that reclaims some of its territory, and must be reckoned with, but also struggles to survive in urban canyons.

The mere publication of this article in *JAPA* is a positive sign. The field of urban planning can be quite conservative, and this is one example of endorsing some important out of the box thinking about nature in cities.

Can we manage parks and greenspaces in cities more ecologically and naturally?

There is a serious opportunity to manage many institutional spaces in and around cities to enhance biological complexity. We can begin to shift the management of spaces around and between buildings to include not simply standalone trees but rather as more complex, multilayered forests.

How to start this is an open question. Perhaps taking small steps will be possible as they do not challenge the prevailing aesthetic of these spaces. At my own university efforts are being made to allow leaves to collect in the spaces around large trees, creating a kind of expanded ecological zone. It creates a different look and perhaps it will take some time to fully appreciate and accept a wilder, more insurgent urban landscape. Creating these kinds of understories and beginning to think in terms of assemblages of native trees and plants that occur together will be a challenge but an exciting new direction for cities.

Thinking about Water for Trees from the Beginning

Keeping trees alive, especially those young trees during the first three years or so after planting, often raises difficult questions about how and when to water, and especially in more arid cities, where the water for trees will come from. Certainly, compared with provision of potable water for drinking, watering trees is usually viewed as a secondary importance. This should change of course as we begin to see trees as essential elements of a healthy city. Trees that provide shade and address the serious health risks connected to urban heat, must be seen as a priority for use of water,

The strategies used by cities vary greatly, and as drought and scarcity become more endemic in many cities, there are a number of new and creative ideas cities are using. An important initial point is that it will be difficult for a city to meet its tree watering needs without more fully addressing its broader water demand. Especially in arid cities, such as in the American west, there will be immediate perceived conflicts between water use for trees and water uses such as watering lawns and provision of drinking water.

The first and perhaps most important set of strategies will involve choice about which trees to plant, and in an arid climate native species that are already well adapted to low water conditions should be the trees of choice. In cities like Tucson and Phoenix, for example, native species such as velvet mesquite, blue palo verdes, and desert ironwood. These are species that have evolved in this arid landscape and can usually survive for long periods without watering. The palos verdes has evolved a series of physiological strategies for adapting to drought, for example, including the ability to photosynthesize through its green bark and to drop its leaves when necessary.

In Tucson, an initiative called SOMBRA[19] is aiming to reintroduce velvet mesquite trees to underserved neighborhoods, especially in the south of the city, where tree canopies are lower and heat danger higher. With a grant from the Arizona Department of Forestry and Fire Management, the Community FoodBank of Southern Arizona is building a series of "shade huts" around the city, where mesquite saplings will grow, eventually to be planted in neighborhoods. The seed pods of the mesquites can be harvested and ground into mesquite flour, so part of the goal of this initiative is also to address food insecurity in these neighborhoods.

Water conservation in such cities will also have to be a major part of the answer. There is no reason why precious water should be used to sustain biologically sterile turfgrass lawns for example, and many cities from Los Angeles to Las Vegas have adopted financial incentive programs to encourage the removal of such water-intensive yards and their replacement with "xeriscaping." Often referred to as "cash for grass," these programs recognize the importance of conservation as a first step in addressing conflicting water demands.

Other ideas include the creation of onsite green streets and green alley initiatives that seek at once to take out impervious surfaces and to create bioswales and linear rain gardens that capture stormwater and allow it to percolate. Such green street designs have been extensively used in cities such as Portland, Oregon, and are just beginning to receive attention in water-strapped Los Angeles. What makes sense is to begin to plan such green streets to include trees as an integral park, not only helping to replenish underlying aquifers but also to steer stormwater to the root structures of thirsty trees.

Tucson is now commonly designing tree planting in combination with rainwater collection. Its green infrastructure initiative, called "Storm to Shade," explicitly links expansion of its canopy—and it has adopted a citywide goal of planting one million trees by 2023—with the use of stormwater to sustain them. Water and trees must necessarily go together in this extremely arid city.[20] A new Green Stormwater Fee (which costs residents around $1 dollar every month) funds these projects: they include planting trees in chicanes that narrow roadways and actually steer stormwater to the trees. The before and after images of these planted chicanes are impressive—the small amount of rainwater that falls on the streets of Tucson is more than enough, if not wasted, but instead steered and captured (it legalized curb cuts that allowed water to drain to these planted areas in 2007).[21]

A comprehensive approach to water means looking for ways to capture and use all sources—this means not only rainwater and stormwater but also reuse of greywater, from a building's kitchens and bathrooms. This minimally contaminated water is perfect for plant and tree watering but in most cases is wasted by being sent down a drain.

LA is starting to support efforts to reduce impervious surfaces and allow more water to infiltrate and replenish its groundwater supplies and this will ultimately be good for trees. One example is a new green street pilot in the Pacoima neighborhood of Los Angeles has been under development. The Green Street on Laurel Canyon Drive captures, retains and allows the infiltration of an estimated more than 13 million gallons of stormwater each year through a vegetated bioswale that adds greenery to the neighborhood and also helps reduce local flooding.[22]

For a city like Los Angeles billions of gallons of water is lost through drainage to the ocean (creating itself marine contamination issues). Larger facilities like Santa Monica's SMURRF (A wonderful acronym standing for Santa Monica Urban Runoff Recycling Facility) are another part of the answer. This is a unique kind of urban infrastructure—a facility that collects and filters runoff from a large swath of the City of Santa Monica, processing up to 500,000 gallons a day of very clean water that can be used for landscape watering and other non-potable uses. The facility is also located in a spot (close to the popular Santa Monica pier) that makes contact with the public easier—indeed the SMURRF is designed in a way to educate the public, including design that makes it possible to walk through it and places where the treated water is intentionally made visible. There are "overlooks" to see and learn about the different steps in the treatment process as well as artistic elements—colorful tile and lighting—that also make it an interesting visit.[23] How such a facility can help to ensure the survival of the city's trees and canopy perhaps ought to be a more prominent part of the educational story here.

Many water-deficient cities are looking critically at ways to adjust their urban landscapes to capture and retain more rainwater for eventual potable purposes. In Los Angeles, there has been considerable criticism for the failure, despite goals to the contrary, of taking steps to de-seal its largely nonpermeable streets and built environment. Recent extreme storms and flooding events there has reinforced this since of missed opportunity—watching engineered flood control structures (like the LA River) speedily deliver water to ocean outfalls, at the same time as it invests in expensive and damaging technologies such as desalination suggests the folly of their approach. Trees and forests that capture and retain storm and flood waters can be much of the answer actually, allowing water to seep in and to recharge the city's groundwater sources.

More buildings are now being designed to capture the extensive condensate from their air conditioning units, and in cities such as Austin, San Antonio, and San Francisco there are now significant financial incentives to encourage this. Austin's new Central Library is an excellent example. A comprehensive water design for these buildings—including capture of rainwater, use of reclaimed water and harvesting of condensate, provides some 90% of the building's water needs. Much of this recycled water is used to water the building's vegetation and landscaping. This includes trees planted on the library's roof garden, which includes a live oak tree.

In Singapore there are some impressive new biophilic buildings that show what is possible through comprehensive water design. These include the WOHA-designed Kampung Admiralty, where stormwater is collected from its largely forested, multi-tiered rooftop and used for landscape watering.

Just as new buildings will need to think more comprehensively about water use, perhaps we need a model of new development that better incorporates trees and tree planting, with the watering needs of those trees thought of and incorporated in the design from the beginning. We do have examples—for instance the Pearl Brewery in San Antonio harvests its condensate and uses this to irrigate an adjacent vegetated courtyard. The water needs of plants and trees onsite should be a regular consideration in the design and approval of major new building projects in any city in the future.

The Role of Tiny Forests?

Is it possible that many small spaces in cities represent opportunities to plant forests that are more than a tree or two, and perhaps the chance to install diverse, multilayered mini forests? This is the essential idea and innovation of the Miyawaki forest that has begun to gain momentum in cities around the world. Named for its inventor, Japanese botanist Akiro Miyawaki, these mini, or tiny forests as they are sometimes called, have many advantages.[24] While they are not designed or intended to lead to large amounts of forested land in cities, they can be an especially effective way of engaging the public. And they could, if planted *en masse*, have a meaningful, cumulative impact in cities, though the extent of these effects remain to be seen.

I first learned about these, as many have, by watching one of the many TED talks online. Especially impactful has been the work of an Indian environmental activist Shubhendu Sharma, whose TED talks are quite compelling. Sharma started a now successful nonprofit called *Afforestt* to put these ideas into practice there in India.

I had the chance to talk with Sharma by Zoom several years after I had first heard about him and his passion and compelling message were still quite obvious.[25] He described his remarkable efforts helping many organizations to plant forests. He spoke of the need for a shift in our urban mindset away from seeing beauty in a sterile lawn to seeing the beauty of wild nature. In the beginning of our conversation, he showed several images of the Taj Mahal: comparing a current image with few trees, with a historic image showing the iconic structure surrounded by a wild forest.

Sharma described a recent forest project in New Delhi in which wildness and biodiversity have returned in a short two years since the forest was planted. "Birds, bees and butterflies are coming back," he told me, and they have seen a remarkable reduction in temperature there compared to the surrounding urban spaces. The next step he believes is finding ways to integrate human activities (something he is calling "forestscaping"), something he is working on with the Delhi forest, so that it might be possible, for instance, walk and sit and even to work from the forest.

What is different about the tiny forest? It is a method and model that entails extensive planting of native species of forests. Not spaced apart, but planted very close together, in dense layers. The intent is for these trees and shrubs to grow very quickly and require relatively little maintenance. Some watering and weeding will be necessary in the first couple of years, but they are designed to be self-sufficient for the most part after that.

Sharma has inspired many other adherents and practitioners of the tiny forest around the world. One of them, Nicolas de Brabandère, based in Belgium who, after seeing the TED talks, found himself traveling to India to learn more. Today he runs a for-profit business, called *Urban Forests*. He recently told me by phone that he has, in six years of work, helped to plant some sixty tiny forests. The work seems to be speeding up for his company, telling me that he completed nineteen forests in the last year alone. And his small company has now grown to include six employees (demonstrating in a small way the jobs and employment potential of planting trees!)

I asked de Brabandère about the lessons he's learned from planting so many forests.[26] Some of them are technical—for instance, the need to pay

FIGURE 4.2 Nicolas de Brabandère Standing in front of One of His Tiny Forests. Photo Credit: Nicolas de Brabandère.

careful attention to preparing the soil, which is a key step in the Miyawaki method (Figure 4.2). Only three of the sixty forests he's planted would be considered failures, he notes, and in one of these cases because he did not pay enough attention to preparing the soil. Another lesson is to ensure that clients understand what the resulting forest will look like. It will be different, maybe to some not the "clean" and tidy forest, but a quite messy looking one.

The sizes of the tiny forests he has planted have varied, from only 100 square meters to as large as 3000 square meters. He believes the latter is as large as you should go with the Miyawaki method. Any larger would really require a different approach or method.

"The Miyawaki is not for reforesting the planet," he told me: "it's about creating something special that creates an emotion in people's hearts so that they want more." These tiny forests are portals; ways to engage and educate the larger public and to convince about the larger need to conserve and grow more trees and forests. At several points in our conversation, he refers to their "transformative" value. "It's really a seed in people's minds," he notes.

And the process of planting, and for the first couple of years maintaining, the tiny forests often involves citizens directly. "It brings people together," says de Brabandère, and is something residents create and are proud of. It also improves the quality of the surrounding neighborhoods, and often helps for instance to create a natural sound barrier ro traffic noise.

Expectations about the carbon-sequestering potential of the Miyawaki forest should especially be tempered. They are not likely, at least in any short timeframe, to help much in locking up carbon. Protecting our larger trees in cities will be a more productive approach. But they can do much to quickly change the ecology of a site, to repair and repopulate the biodiversity lost from cities. And Miyawaki forests are popping up all around the world showing considerable promise in bringing nature back to otherwise degraded and sterile urban sites. Like the Miyawaki forest in Beirut planted along the Beirut River, a highly polluted concrete conveyance structure. Spearheaded by a local architect and biomimicry adherent Abid Dada, and helped by Afforestt's Sharma, it is a small parcel (2200 square meters) but home to nearly 2700 trees, and many shrubs as well, planted mostly by volunteers. The mix of native trees was informed by a trip outside the city to a stretch of more natural river upstream.[27] This small project is already showing signs of life returning, even just after two years: "So many lizards and geckos and we're observing the return of many birds," says a two-year progress report.[28]

The Netherlands has emerged as an epicenter of tiny forest activity, largely through the work of the NGO, IVN. Daan Bleichrodt runs the tiny forest program there and in June 2023, we visited together several tiny forests in Utrecht.[29] They have impressively been able to scale up, with nearly 200 tiny forests completed in public spaces of cities and towns throughout the country. Their model is to work collaboratively, with the local community and usually with a specific nearby school. Here students are engaged in planning, planting and maintaining the forests, and can become "forest rangers." The typical cost for a forest is around $20,000 Euros, about half of that for the plants and trees. The costs have been shared, with IVN covering half and municipalities typically covering the other half. In 2018, IVN received a 1.85 million Euros grant from the Dutch postcode lottery, which has helped greatly to scale up the forests and to give visibility and credibility to the initiative, a "stamp of approval, Bleichrodt says. Bleichrodt tells me the tiny forests are now in seventy municipalities, and in some cities there are multiple forests (Almere has planted ten!) Another 100 smaller forests have been planted in the private garden spaces around peoples' homes, and creatively, Bleichrodt and his colleagues have even started selling a "forest in a box," online, which arrives with bare-root trees and shrubs and instructions in how to plant them (they have already sold 5000 of these!)Figure 4.3

FIGURE 4.3 Daan Bleichrodt, who runs the tiny forest program at the NGO IVN stands in front of one of the tiny forests in the City of Utrecht. Photo credit: Tim Beatley.

The Dutch tiny forests have a target size of about 200 square meters but can be larger. One forest I visited, the tiny forest at the Muziekplein just a few stops from the Utrecht Terwijde Train Station, was 400 square meters in size, and included more than 1100 trees, and around thirty-five different species. Like all of the Dutch tiny forests it is enclosed by an attractive wood and wire fence, with a pathway through the middle. Each forest also includes an area of public seating, often used as an outdoor classroom. Especially intriguing about the Muziekplein forest is what it replaced: twenty car parking spaces. The Dutch tiny forests have been studied by researchers at Wageningen University and the results are positive—finding considerable biodiversity, in one case finding 450 species of animals and plants there.

When I asked de Brabandère about the biodiversity of these small forests he recalled an experience from his time in India. He told me that one evening he went to visit a mini forest and was struck by the sounds emanating from it, and the life it clearly harbored. "As I walked to the forest, I felt really good," he said, noting not only how the sounds improved his mood but also just how many creatures were being supported by the little amount of forest space there.

While tiny forests like this are not likely to make much of an impact in sequestering carbon, for example, I wonder if we are perhaps underestimating their potential impact when it comes to local biodiversity. It really depends of course on the scale at which we apply these tiny forests and whether in cities it might be possible to imagine a network of tiny Miyawaki forests that serve as ecological bridges between larger parks and forest preserves in and around cities.

De Brabandère at one point in our conversation describes how he sometimes explains these forests to future clients as being like the small refugia that existed during the last ice age. How important could they be in passing along and supporting biodiversity over time? And could these small refugia be designed in ways that allow them to serve as seed banks and mechanisms for repopulating the non-forested spaces around them. In many American cities one concern will be whether newly planted tiny forests are eaten or overgrazed by deer.

This reminds me of my visit to a very unusual but inspiring forested nature preserve in Wellington, New Zealand. Called *Zealandia*, it is one of the largest examples of an effort to create spaces to allow native birds to rebound.[30] New Zealand's native birds have been decimated by nonnative species like weasels, stouts and of course domestic cats. Zealandia's main conservation strategy is the erection of a predator-proof fence that provides safe space for these native birds to reproduce and thrive, which they have. Zealandia has been a success story as residents in neighborhoods surrounding the preserve (and now most parts of the city) have begun to hear and see birds they hadn't seen in a long time.

In many cities native species would benefit equally from such larger protected refugia, including fencing, temporary or permanent, to protect against deer and cats. But I also wonder if tiny Miyawaki forests, or at least some of them, could incorporate such fencing or other physical barriers so that they might protect birds in a way similar to Zealandia.

Conclusions

The management and care of the urban canopy is complex and multifaceted, but incredibly important. And management practices must change, and are changing, in response both to changing environmental conditions (e.g., climate change) and changing perceptions and preferences about trees and forests. Key questions include what species should cities plant primarily and should the mix better reflect the priorities of resilience and biodiversity conservation. We will need to draw more heavily on native species especially in our planting palette, but tree species that will more likely thrive in the hotter and more water-deficient cities of the future. Another key question is how and in what ways trees and forests reflect a desire for wildness. There seems to be a growing consensus that we can and must reimagine our urban forests. Not simply planting and maintaining rows of singular streets, we should pursue more creative planting schemes that might see forest groves extending into streets and up the sides of buildings. We should also make room for processes of decomposition and death, as well as renewal, in urban forests.

Many of the elements of these tree and forest futures will require us to cultivate a different set of urban forest aesthetics, for instance recognizing the beauty and habitat value of snag trees or fallen trees. Or recognizing the beauty of tiny but densely planted forests that, by necessity, are planted in the small and fragmented spaces of cities.

Notes

1 Manuel Esperon-Rodriguez, et al., "Climate Change Increases Global Risk to Urban Forests," *Nature Climate Change*, Vol. 12, October, 2022, 950-955, found here: https://www.nature.com/articles/s41558-022-01465-8, accessed May 22, 2023.
2 Harry Stevens, "Trees Are Moving North from Global Warming. Look Up How Your City Could Change," *Washington Post*, April 26, 2023, https://www.washingtonpost.com/climate-environment/interactive/2023/tree-species-climate-change-north-shift/, accessed May 22, 2023.
3 See Oregon State University, "Principles of Fire-Resistant Forests," found here: https://catalog.extension.oregonstate.edu/sites/catalog/files/project/supplemental/pnw618/pnw618-chapter2.pdf, accessed June 22, 2023.

4 See Pacific Horticulture, Dave Egbert, "Trees in the Fire-Safe Landscape," found here: https://pacifichorticulture.org/articles/trees-in-the-fire-safe-landscape/#:~:text=Deciduous%20trees%20are%20considered%20even,little %20purchase%20in%20their%20canopy.&text=Coast%20live%20oak %20(Quercus%20agrifolia,multi%2Dtasker%20in%20the%20landscape, accessed June 22, 2023.

5 Erik Lind, Thomas Prade, Johanna Sjöman Deak, Anna Levinsson and Henrik Sjöman, "How Green Is an Urban Tree," *Frontiers in Sustainable Cities*, 2023.

6 Douglas Tallamy, *Nature's Best Home*, 2019.

7 Tallamy, 2019.

8 Tallamy, 2019, p.139.

9 Douglas Tallamy, *The Nature of Oaks: The Rich Ecology of Our Most Essential Native Trees*, Timber Press, 2021.

10 Tallamy, 2019, p.63.

11 Tallamy, 2019, p.73.

12 Dengkai Chi et al, "Residential Exposure to Urban Trees and Medication Sales for Mood Disorders and Cardiovascular Disease in Brussels, Belgium: An Ecological Study," *Environmental Health Perspectives*, Vol. 120, No. 4, May 11, 2022, found here: https://ehp.niehs.nih.gov/doi/10.1289/EHP9924, accessed June 4, 2022.

13 Geoffrey H. Donovan, et al., "The Association Between Tree Planting and Mortality: A Natural Experiment and Cost-Benefit Analysis," *Environment International*, 2022.

14 Ibid.

15 See Suzanne Simard, *Finding the Mother Tree: Discovering the Wisdom of the Forest*, Vintage, 2022.

16 Ben Johnson, "Queen Elizabeth's Oak," found here: https://www.historic-uk. com/HistoryMagazine/DestinationsUK/Queen-Elizabeths-Oak/, accessed June 7, 2022.

17 Interview with Bill Cunningham, City of Portland, May 26, 2022.

18 Lucie A. Laurian, Ernest Sternberg, and Nadia Voigt da Mata, "The Transgressive Urban Forest: An Ecological Aesthetic for the Anthropocene," *Journal of the American Planning Association*, 2021.

19 The acronym stands for: Sonora Mesquite Barrio Restoration Alliance, see Caitlin Schmidt, "Trees Project to Bring Shade, Food Sources to Tucson Neighborhoods," August 17, 2022, found here: https://tucson.com/news/local/ trees-project-to-bring-shade-food-sources-to-tucson-neighborhoods/article_ a22339ba-03b3-11ed-a4b1-674b8a6724f3.html, accessed April 10, 2023.

20 See City of Tucson, *Storm to Shade: Pilot Program Report*, November, 2022, found here: https://climateaction.tucsonaz.gov/pages/gsi, accessed April 10, 2023.

21 See Brad Lancaster, "Before and After Photos of Green Infrastructure in Dunbar/Spring," January 11, 2022, found here: https://dunbarspringneighbor hoodforesters.org/2022/01/before-after-photos-of-green-infrastructure-in-dunbar-spring/, accessed April 10, 2023.

22 See "LAUREL CANYON GREEN STREET," LA Environment and Sanitation, found here: https://www.lacitysan.org/san/faces/home/portal/s-lsh-wwd/s-lsh-wwd-wp/s-lsh-wwd-wp-gi/s-lsh-wwd-wp-gi-gs/s-lsh-wwd-wp-gi-gs-lcbgs?_ afrLoop=4530898816366904&_afrWindowMode=0&_afrWindowId=null&_ adf.ctrl-state=50piwvijg_78#!%40%40%3F_afrWindowId%3Dnull%26_ afrLoop%3D4530898816366904%26_afrWindowMode%3D0%26_adf.ctrl-state%3D50piwvijg_82, accessed April 10, 2023.

23 See "Santa Monica Urban Runoff Recycling Facility (SMURRF)," Found here: https://publicartarchive.org/art/Santa-Monica-Urban-Runoff-Recycling-Facility-SMUR/51070947, accessed May 18, 2023.

24 To learn more about the Miyawaki forests see Hannah Lewis, *Mini-Forest Revolution: Using the Miyawaki Method to Rapidly Rewild the World*, Chelsea Green, 2022.

25 Interview with Shubhendu Sharma, March 11, 2021.

26 Interview with Nicolas de Brabandère, of the company *Urban Forests*, August 3, 2022.

27 See Society for Ecological Restoration, "Beirut's RiverLESS Forest," found here: https://www.ser-rrc.org/project/beiruts-riverless-forest/

28 SUGi Project, "Beirut's Riverless Forest," found here: https://www.sugiproject.com/projects/beirut-riverless-forest

29 Visit and interview with Daan Bleichrodt, IVN Utrecht, July 3, 2023.

30 More about *Zealandia* can be found in Timothy Beatley, *The Bird-Friendly City*, Island Press, 2020.

5

FOREST ARCHITECTURE AND DESIGN

It is increasingly likely that in cities of the future we will find trees and forests in places we do not find them today or find them infrequently. They will increasingly make their way into the design of buildings and built form, what I will broadly call "forest architecture." Partly this will be a result of a growing desire for trees and forests in the cities where we live and limited space in which to grow them in the conventional way—in parks, along streets, in urban woodlots. It will partly be a result of the growing recognition of the importance of biophilia and biophilic design and that we want to live in spaces, including the interior spaces of our homes and offices, that connect us to actual living nature (perhaps by bringing living trees inside) or that remind of the nature outside.

There are at least four key ways that urban architecture and design can emphasize trees and forests. One is to design our buildings and cities to reference trees—their shapes and forms. We have done this for centuries, with tree-shaped columns and vaults, and other expressions of dendriform design seen in medieval cathedrals as well as more contemporary designs. Another way can be seen in the growing importance of wood as a building material, partly a reflection of not only how we cherish the look and feel of wood, and its biophilic qualities, but also a recognition of the climate and environmental benefits of wood. We see this in the trend toward mass timber structures (discussed later) and in the development of CLT (cross-laminated timber) technologies. A third way, growing in importance, is the incorporation of living trees into the spaces of a building—on balconies and terraces and rooftops. Trees and even entire forests are being planted in the interior spaces of structures.

DOI: 10.4324/9781003377344-5

There are then many different forms or flavors of forest architecture, and increasingly the blending together of these different ideas in building projects in cities.

Urban Development That Saves Trees and Forests

There is an understandable tendency to think that forest architecture is something new, and it is certainly true that projects like Bosco Verticale showed the world something quite different.

Not long ago I was asked to write an article for the *Frank Lloyd Wright Journal* as part of a special issue on biophilic design. In the course of better acquainting myself with Wright's larger body of work I discovered several projects, early in the modern history of forest architecture (though relatively late in his career) that I did not know about and that surprised me. One was a completed high-rise residential tower in Bartlesville, Oklahoma. Called Price Tower it was his only built high-rise design, and the design takes its inspiration from trees. In fact, Wright referred to it as "the tree that escaped the forest." This 1956 building employs a trunk and branch system allowing the floors to be cantilevered off the structure's central core where the elevators were located. Lily Cao, writing in *Arch Daily* notes that even the building's outer walls continue the tree-references with "long copper wings, their textured green patina suggestive of leaves. Supposedly, the asymmetrical design looks differently from every angle, and in that way resembles the natural but all-the-more beautiful imperfection of a tree."[1]

But most impressive and interesting was the story of a large plan for essentially a self-contained city to be built in northwest Washington, DC. The project was known then as Crystal City (not to be confused with the neighborhood by the same name in Arlington County, across the Potomac River) or Crystal Heights. Designed during the summer and fall of 1940, Wright envisioned it as a series of high-rise towers, aligned along a crescent, and focused upon an existing, old grove of trees. This grove included the so-called Treaty Oak, a tree that was believed to be between 350–400 years old.

The grove of trees became the centerpiece of his design. It would have been massive in scope—a residential-hotel complex, consisting of two dozen towers, each twelve to fourteen stories tall. There was to be a high-end shopping mall, a movie theater, and a multistory parking garage. It would have been, for the time anyway, "the largest apartment-hotel in the world."[2] Had it been approved and built it would have been, in Wright's words, a "city within a city." "The Heights will mostly be seen as slender glass obelisks rising through the mass of big trees culminating in the white marble shafts of the ventilating system ..."[3]

It was to be located on one of the city's last remaining natural sites and remnant forest. Connecticut Avenue on one side (to the west) and the existing Wyoming apartments to the north, today the site contains a Hilton Hotel and nothing remains of the forest that was there. Wright's design would have preserved the oak grove and clustered the high-rise buildings in an arc around the forest, making it the centerpiece. Wright's client, a shady developer, pushed Wright to increase the density of the project, something Wright resisted, aiming to protect as much of the natural site as possible.

Neil Levine in his wonderful (and very comprehensive) book *The Urbanism of Frank Lloyd Wright* describes Wright's intent to set half the site aside for nature: "nearly half of the upper plateau is devoted to a garden preserving much of the original oak grove, with the tower grouping essentially forming a frame for it."[4] There are several renderings of the project that survive, showing as Levine notes, "The hotel and apartment towers form a continuous, serrated, amphitheatric backdrop to the oak garden in its embrace." Use of the word *embrace* is appropriate, and the design does seem to show the towers cradling the trees.

Here we find an early example in urban planning of a project designed to start with protecting the trees in the city, clustering and orienting buildings to steer clear of them, and going even further than this—indeed making the trees and oak forest the key focal point for this not insignificant community (of some 2500 units in total it should be remembered) who would have been living here.

Crystal Heights was ultimately nixed by the DC Zoning board, which rigidly enforced the city's famous building height limitations, dating back to the Height of Buildings Act of 1899. Wright's organic architecture and iconic projects like Fallingwater, perhaps the most famous biophilic building, have cemented his credentials as a designer who put nature at the foreground. But he is not often thought of as a green urbanist. In this design he shows the prospect of supporting considerable urban density and vertical urban form but at the same time making room for trees and forests. The result is a sad one, in a city that values trees, and the project had it been built might have demonstrated how special and valuable the trees were, and an enhancement to development at any scale. I would certainly have loved to have seen and touched the Treaty Oak.

There are still relatively few good examples of recent development projects in cities that have protected existing older trees and show that it is not only possible to build and develop but also to save the urban forest. In a recent conversation with the forest conservation group *The Last 6000*, in Seattle, sadly few examples came up from that group of development projects protecting and integrating trees; they could simply not think of many. One that did come up was a new residential project called Bryant Heights. Completed in 2018, and designed by Johnston Architects, it is a project that mixes single family, townhomes and "a mixed-use building that extends the commercial node along NE 65th St," in the Ravenna-Bryant neighborhood. The design of the project managed to save twenty-five existing trees and incorporates a "cluster of Douglas Fir trees."

In a 2018 Seattle Times article entitled "5 Architectural Approaches That Are Shaping the Way We Live," the Bryant Heights development is heavily profiled as an example of new multifamily housing sensitively fitting into existing single family areas.[5] Described in the article as "thoughtfully transitioned, gracefully scaled." the design and buildings are "all purposefully centered on towering old trees, welcoming public pathways and—most important—the fundamental mission of blending into its Ravenna-Bryant neighborhood."

The online advertising material and description of this project highlights the trees to be sure, describing Bryant Heights as an "exciting and unique enclave of new homes ... where old growth trees and lush landscape grace the walkways and open spaces." It is surprising that more developers and builders do not appreciate or understand the potential ways in which trees and designing with trees from the start could enhance their projects and indeed enhance their desirability and profitability. The protection and integration of trees, moreover, represents one of the most important things a designer or developer can do in fitting a project into an existing neighborhood. There is the positive trend in many cities, including Seattle, of increasing permissible densities and finding creative ways to fit more units into existing single family neighborhoods.

Projects like Bryant Heights show how essential trees and nature can be. They will soften the look and feel of multifamily projects and they will help residents of existing neighborhoods see the value of new denser forms of development. In the case of Bryant Heights the trees and the site had been informally used by the neighborhood as a park, and saving the trees amounts to an act of being respectful to the neighborhood and its character, while offering those living nearby some significant nature amenities.

As architect of the project Mary Johnston notes, "The trees really help the across-the-street neighbors. It feels much older with the mature landscape. From a practical point of view, a project like this one is likely to see less vocal opposition, likely to gain approval faster, likely to diffuse or overcome the often-vociferous NIMBYism seen in response to denser projects."

Architect Mary Johnston explains the benefits this way: "Leaving a bunch of trees, opening it up and inviting neighbors to walk around and enjoy it—that shows how you can integrate multi-family into a mostly single-family neighborhood, and have a successful project that people like."[6]

Do we need to reach out to and work to educate the development community about trees and urban forests and the essential role they can play in protecting and expanding the urban canopy? Probably so.

Trees will continue to be an important element in many of the new designs for homes and buildings that will be necessary as we begin to tackle climate change in earnest. There are now a growing number of home and office designs that seek to dramatically reduce the consumption of energy and the emission of carbon.

Another stand-out development where a remarkable effort to protect and integrate older trees can be seen is Oak Terrace Preserve, a part of the larger redevelopment of the Charleston Naval Base (in North Charleston, South Carolina) called Noisette. This neighborhood of single family and townhomes (300 and seventy-four units respectively) began with an extensive inventory of the trees on site (Figure 5.1). Efforts were made at every stage of development and construction to protect the more than 600 trees, many that would be considered "old growth" and larger than 24 inches DBH. The Center for Watershed Protection has prepared a thorough case study of this "forest friendly development."[7]

Years ago I traveled to North Charleston to better understand Noisette, and the sustainability vision of its ambitious and very forward-looking developer John Knott. I took photos of the homes that had been built in Oak Terrace Preserve, and walking around this neighborhood it had the feel of an older place, where the trees and homes blended together. It was clear that this tree-centric neighborhood was going to be an unusual place to live, and one that allowed its residents to feel closer to nature and a sense of immersion, living in a coastal forest. The beauty those trees added was undeniable.Figure 5.1

The Center for Watershed Protection's case write up makes prominent mention of the residents experiences:

Oak Terrace Preserve has proved to be an extremely popular location for people to live due to its tree preservation efforts. A poll of residents found that the presence of the trees in the development was one of the top three reasons for selecting to buy at Oak Terrace Preserve. Homeowners in the development have also taken on the role of tree protection advocates themselves, often reporting to the Noisette Company if they believe that current construction is encroaching unnecessarily on trees in the development.[8]

In this development the trees were understood as a key asset and their protection and presence became the central design factor. The importance of tree protection fit well within and clearly followed the larger ecological

FIGURE 5.1 At Oak Terrace Preserve, part of the development Noisette, in North Charleston, SC, the homes are placed in ways that preserve the existing trees. Photo credit: Tim Beatley.

priorities and ethics set by Knott and the larger redevelopment project. And at every stage along the way, including during construction, the health and protection of the trees was given priority.

In the summer of 2022, I visited one of the most interesting examples of a new generation of zero-energy projects., this one in the City of Arvada, north of Denver, Colorado. Called *Geos*, it is the brainchild of Austrian engineer Norbert Kelb. Its aim is to dramatically reduce energy needs and carbon emissions. The Geos website describes the project as one that "combines traditional village living with the most advanced design and building practices. Community members enjoy a pedestrian lifestyle with front porches, tree-lined sidewalks, corner stores, and neighborhoods."[9]

It has been dubbed the first net-zero-energy neighborhood in Colorado. "Geos utilizes the sun and earth to produce as much energy as it consumes, with a total cost that equals to or less than ordinary built-to-code communities."[10] The development is relatively dense with buildings sited close together, but with an interesting twist: the homes are staggered slightly to ensure that no home shades another. The neighborhood's design allows the homes to orient to the south, to maximize the capture of solar energy, both passive gains and photovoltaic panels.

In combination with other technologies and low energy products—use of ground source heat pumps, low energy appliances, extra levels of home insulation, and triple-pane windows—it's estimated the homes cut energy use by 75% over more typical homes. The strategic placement of trees plays an important role in this design: south facing windows capture sunlight in the window and deciduous trees provide shade in the summer. "During the darkest days of winter when the sun is lowest, all south facing windows receive full sunlight to help heat the home. In the hottest days of summer, overhangs and deciduous trees prevent the sun from hitting any windows directly, keeping the homes naturally cool."[11]

As these examples from Seattle and Arvada show, trees can be an important and highly desirable (and desired) feature to home buyers and ought presumably to be important for this reason to developers. Yet the research (and my personal impressions) suggests that trees are not getting the attention and support by the real estate profession that they need and deserve.

Especially in response to the stresses of Covid, there has been lots of discussion about what if anything has changed, or ought to change in home design in the future. Have homeowners shifted in their preferences for what kinds of houses and living environments they want in response to the experiences of COVID? There are some good clues about this. One clear message is that connections with nature have been important during COVID. Nature in all its forms have been salve and a balm and have helped urban residents sustain themselves during lockdowns. More and more people, for example, have discovered birds, and are bird watching and listening to birds.

Not surprisingly, surveys that aim to understand trends in home design are finding this new attention to nature. One recent survey of forty-seven "homebuilders, designers and architects" asking which features were most important to homeowners in 2022, found the number one feature to be "usable outdoor area." This beat out other important features, such as natural light and home offices, that were also highly ranked, as well as a number of other features such as solar panels and smart features that were ranked much lower.

What does this mean for trees and the perceived need for trees in residential settings? For many of us the idea of spending time in the outside spaces around our homes is intimately connected to trees. Interestingly, the photos used in the article to accompany the results were of spaces where trees and shade are prominent elements. These images of backyard spaces seem especially desirable, as the magazine photos show. There is also considerable evidence that the free market rewards the presence of trees in the sale of homes. In an important study more than a decade ago, researchers found that proximity to street trees in Portland, Oregon, added almost $13,000 the sales price of homes, and also shortened the number of days homes were on the market.[12] There are clear economic and market advantages that trees deliver and important ways in which protection of existing urban trees will improve the salability of a home or development.

Nevertheless, it seems that developers and homebuilders tend to undervalue trees and often do not hesitate to remove them when inconvenient, without, it seems, much thought about what that does to the desirability or market value of these home products.

Integrating Living Trees into Building Design

Italian architect Stefano Boeri has perhaps done more than any designer to help us reimagine how living trees can be an integral part of a building or development. His pioneering project is *Bosco Verticale*, a pair of residential towers with exterior spaces designed for planting trees and plants. They are a remarkable sight and a project that lives up to the beautiful renderings many of us admired ahead of its construction. It does indeed look like and function as a "vertical forest," as Boeri refers to it. Some 800 trees and thousands of plants and shrubs lovingly embrace the living spaces here. As the Boeri website describes them, "the towers are mainly characterized by large, staggered and overhanging balconies (each about three meters), designed to accommodate large external tubs for vegetation and to allow the growth of larger trees without hindrance, even over the three floors of the building."[13]

As fascinated as most of us were when the design for Bosco Verticale was unveiled, Italians might have remembered that this idea originated hundreds of years ago in their country. Sometimes famously described as the first ecological rooftop, there is the remarkable story of the Guinigi Tower in the Tuscan city of Lucca. Built by the wealthy Guinigi family in 1384, it boasts a small oak forest on the top of this 125-foot structure, one of some 250 towers in the medieval period meant to serve as a demonstration of wealth and power. There are at least five large trees reported to be holm oaks that are 100 years old.[14] This "oak tree garden," as it's called in some publications, is a visually striking element of vertical nature and must surely have been ideas to contemporary designers like Boeri (I wonder if he has visited? I suspect, yes).

Making room for large trees many floors up is no small feat, of course. Interestingly, the occupants of flats at Bosco Verticale are likely to be seeing and experiencing the trees that are growing skyward from three stories below. The trees are grown in a special nursery and are braced to ensure they are stable during windy times There is a team of "arborist-climbers," or as Boeri calls them the "Flying Gardeners," who prune the trees yearly.[15]

Other forested tower projects have been inspired by Boeri's work and have even sought to improve upon it. For instance, Designer's Walk in Toronto. I spoke several years ago with Brian Brisbin of Brisbin Brook Beynon Architects, who explained the level of design detail that has gone into this project. The incorporation of the trees has been described as "plug and play," with even the building's floor plates designed to include growing space for the trees.

One key lesson from this project is just how much an urban neighborhood can benefit from and appreciate trees even when they are on someone else's balcony or terrace and how they can contribute to the beauty of the surrounding neighborhood. Brisbin explained to me that while the project received some resistance from the city's planning office, there was strong support from the neighborhood. This support he believes helped ultimately in gaining approval. This story is contrary to the typical NIMBY dynamic (standing for "Not in My Backyard"). Proposals for tall buildings and denser forms of development are often met by vociferous opposition in current residents' lower-density neighborhoods, often dooming a project. What the Toronto story suggests is that the trees and green terraces worked not only to reduce opposition but also helped residents to see the project as a positive addition to the neighborhood. At a time when we need to densify, for environmental and equity issues, trees and other natural design elements will be an important tool in overcoming "NIMBY-ism."

Vietnamese architect Vo Trong Nghia has made an international name for himself in designing homes and buildings that include nature, and especially trees. Some of these projects are quite unusual, startling even. The firm will only take clients with a similar embrace of nature. "I love trees and forests," he told an interviewer of *Stir World*, "I always dreamed about living in a house that would feel like being in the middle of a forest."[16] One important example of this work is his aptly named House for Trees. It is a very unconventional home designed for a friend experiencing depression, channeling his own desires to live close to trees, the house is uniquely organized as a series of smaller vertical buildings with trees on top of each. Each building has a different function, or "pot" and the occupants must move between them, and in the process spend time outside. This was all intentional to Nghia.

Exterior trees designed into terraces might also serve to create forested walks in the vertical spaces of buildings. Several new designs for urban office towers include connected terraces that snake around the structure rising to its top. One example is The Spiral Tower, designed by BIG architects, and given its location across from the High Line in New York, has floated the intriguing idea that perhaps this now-iconic urban walk could be extended even further, in this case skyward. The newly designed but not yet built Amazon Helix, at its HQ2 campus in Arlington County, Virginia, is another interesting example. The design imagines a public forest spiraling up and around the structure. According to Amazon's John Schoettler, Vice President of Global Real Estate, the Helix will "feature two walkable paths of landscaped terrain that will spiral up the outside of the building, featuring plantings you may find on a hike in the Blue Ridge Mountains of Virginia."[17]

There is a version of this at the Park Royal Hotel in Singapore, where on one level there is a heavily vegetated trail where visitors can take a hike. WOHA-designed structures are remarkable for the trees and nature included—not afterthoughts or architectural details but rather front and center. Kampung Admiralty is one of my favorites. It is a forested tower of a different kind (see Figure 5.2). Almost all of the flat roof space of this eleven-story mixed use project is covered in trees. Looking down from above, as if a bird, it would be hard to tell that it was anything other than a forest. This WOHA-designed project is special in a number of ways: it mixed use design that includes housing for seniors, a health clinic, day care, a food court, and on the ground level a beautiful public plaza, connected by the pedestrian paths and network around it and creatively shaded by the floors above.

FIGURE 5.2 WOHA-Designed Kampung Admiralty Project in Singapore, with a Tropical Forest Growing on a Multi-Tiered Rooftop. Photo Credit: Tim Beatley.

Buildings That Embody the Forest

There are still other ways to think about forest architecture. One is to imagine more buildings made from wood and timber, but ideally sourced locally and/or from sustainable forest operations. Could the sourcing of wood as a building material actually contribute in a meaningful way to the conservation of global forests?

The Bullitt Center in Seattle is an exemplary and pioneering sustainable building and one of the first certified living buildings under the Living Building Challenge. As such there are fairly rigorous performance standards that have to be met, most impressively these buildings are required to show that they will be net-zero-energy and net-zero water, among other things. In the case of the Bullitt Center this has led to some prominent architectural elements. To find sufficient roof space to accommodate the necessary energy production from photovoltaic panels, the roof is extended in dramatic fashion over the adjacent sidewalk. A recent analysis of the energy produced and consumed by the building over the last decade reached an impressive conclusion: that there was almost 30% more energy produced than consumed.[18]

Here there has been a careful strategy for sourcing the building's lumber from more sustainable sources. Specifically, all wood used in the structure is FSC-certified and sourced within a 1000 kilometers (620 miles). Much of the interior is exposed wood, including the structural timbers, as well as the so-called "irresistible stairs," which seeks to nudge physical movement in the building (Figure 5.3). Taking the stairway exposes one not only to the beautiful wood but also treats to a view of downtown Seattle (there is an elevator as well, but is intentionally hidden).

There have been many initiatives in cities, including a number of private companies, aimed at reclaiming and recycling wood and timber from trees that have fallen down in storms or must be removed for other reasons. The resulting furniture is beautiful and often provides a helpful connection to the trees and forests in which the building or home is embedded.

Several years ago, a group of graduate architecture students started an initiative to reclaim and use fallen trees from our campus. Mostly such trees are carried away quickly, deposited in a facilities management yard, often converted into mulch chips. The resulting tables and other furniture created by the students are stunning and can be found throughout Campbell Hall (the School of Architecture).

With all such efforts at reclaiming downed trees I have my worries about too quickly justifying or rationalizing the cutting down of trees in the first place and the failure to allow snag trees and fallen trees to remain where they are. More localized, sustainable wood supplies are an important way

FIGURE 5.3 Wood of the "Irresistible Stairs" of the Bullitt Center, Seattle. Image Credit: Tim Beatley.

in which we can support the emergence of circular metabolisms, and can be an important way to reduce the demand for and consumption of distant trees and forests. We need to recognize some of these difficult tradeoffs, and ensure that in our efforts to support local wood production do not have the unintended result of losing urban trees or important tree-connected biodiversity.

Another certified living building, the Frick Environmental Center, in Pittsburgh, incorporated a theme of trees and forests throughout the building. Much of the material used both inside and out is wood, and entering the structure one immediately notices this, including the front desk and benches throughout. Much of the exterior of the building is clad in Black Locust, chosen for its durability. No paint or staining was required here. The wood panels that form the interior walls of the Center are designed in a way that is meant to mimic the rhythm of a forest. As education director Camila Rivera-Tinsley told us in the making of a short documentary film about the building, the sizes of the wood panels were "irregularly pattern intentionally" (Figure 5.4). "When you look out in nature you don't see cherry trees and pine trees and oak trees all the same size, all spaced two feet apart." Rather, a healthy forest contains a diversity of lengths and ages of trees, and the geometry of this environmental center seeks to simulate these more natural wood patterns.

FIGURE 5.4 The Wood Exterior (Black Walnut) of the Frick Environmental Center, Pittsburgh, PA. Image Credit: Tim Beatley.

A similarly creative example of a tree- and forest-themed design can be seen in an unusual elementary school built in Manassas Park, Virginia. It was designed by architectural firm VMDO as "a school in the forest," taking full advantage of an existing forest located next to the school site.[19] The school's design maximizes this adjacency, working through how the existing adjacent forest can be drawn in and extended into the school's two main courtyards and outdoor learning spaces. These "forest court-yards," as they are called, are planted with new trees and that will "over time become contiguous with the adjacent forest."[20] The existing forest also becomes a presence in the school through the use of large windows that face it. The school's three different floors are reconceptualized as layers or levels of a forest: the forest floor, understory and canopy, and different sections of the building organized into season-themed "houses." Classrooms are not given numbers but species names. The tree and forest theme for the school is carried throughout the building in other ways as well, for instance through the use of vertical strips of wood along the walls, meant to mimic the irregular tree patterns and slants in a forest.

At an even larger scale is the new main terminal at the Portland, Oregon, international airport, with not only much regionally sourced and sustain-ably harvested timber but also an interior design meant to mimic the feeling of a northwest forest. Designed by ZGF Architects, the structure utilizes some 3.3 billion square feet of Douglas Fir, much of it from the forests of native American tribes in the region. The most dramatic feature will be the structure's mass timber roof. The project's renderings show many actual living trees as well and together will lead to a magical walk under and through as one finds the way to one's airline gate. In the words of ZGT architect Sharon van den Mueller: "The roof design was inspired by the forests of the Pacific Northwest and the feeling you get while walking through the woods, the experience of light filtering through the trees, and the protection of the tree canopy."[21] A trip departing or arriving from this airport will entail the chance for some forest bathing and an opportunity to lower the stress and anxiety levels connected with airline travel.

Many biophilic buildings now also commonly include a variety of nature-based forms of art and often these include trees and forests as an important theme or subject. Hospitals have begun to realize the healing and therapeutic value of such art and have been investing in extensive art collections. The Cleveland Clinic, for example, now has a massive art collection of some 6800 pieces, and one of the most interesting and most popular is a "digital animation" of a tree by artist Jennifer Steinkamp. Called "Mike Kelly 1" (in honor of one of her art teachers), it takes the form of an 8-minute-long projection against a wall. The digital tree moves and shows the life of the tree over four seasons, including the loss of its leaves. This digital tree is one of the art pieces patients and visitors seem most drawn to.[22]

A more recently certified living building is the Kendeda Building for Innovative Sustainable Design, on the campus of Georgia Tech University, in Atlanta. Wood again figures as a prominent building material, much of it in this case sourced from downed trees on campus and milled in a local mill. The building is another example of a certified living building, so is net-zero for energy and water, and also uses bird-friendly glass. The wood found throughout the structure does much to enhance the biophilic interior space of this building. Increasingly cities are turning to tall timber structures and use of cross-laminated timber construction. Such structures have the advantage of avoiding high-energy and high-carbon steel and concrete, and also the ability to provide biophilic interiors that actually sequester carbon, something cities will need to be doing more of. The lessons from Kendeda for shifting the material flows of wood are considerable--the Atlanta-area Mill used (Eutree) describes its approach as "forest free," in that it relies on salvaged wood from the Atlanta metro region.

Equity is a key aspect of this model in that it sends dollars into the local community and is designed for universal access. A portion of the building was constructed by "previously unemployed members of a workforce development program. They moved on from this project with new skills and opportunities."[23]

The positive shift toward the design and construction of buildings with wood holds significant potential as a global forest conservation. Wood is again much less energy and carbon intensive than other building materials and is a renewable resource. It is potentially possible to sequester quite a lot of carbon in wood and timber structures and also to support sustainable, regenerative forestry in source regions and countries.

Projects like the new Portland airport will include a lot of wood and timber and also seek to simulate in the interior spaces the look and feel of forests. Other buildings have sought to design the larger exterior shape of the structure in ways that resemble a tree. Years ago while living in Western Australia I visited a forest products center, where they were exploring new forms of art and craft-based products that would more sustainably use wood from nearby forests. The building was dramatic, assuming the shape of a Yarra tree leaf, though without a bird's-eye perspective it was not apparent.

Imagining a building, especially a tall building, as a tree is not new in the architectural world. One of the best examples can be seen in the philosophy and design work of Frank Lloyd Wright's, especially in his latter years. His organic design philosophy was innovative and ahead of its time in its view that all buildings should blend and integrate into the larger landscape.

Equally valid forms of forest architecture can be seen in buildings and projects that incorporate visual references to trees, if not actual living trees. The Interface Carpet headquarters building in Atlanta, Georgia, is a dramatic case in point. This retrofit of an older structure incorporates a number of biophilic features but its exterior facade, a life-size eastern US forest, is its most dramatic feature. "The most striking feature," describes Perkins and Will, the architecture firm that designed it, "is its skin: 307 panels of glass wrapped in a semi-transparent, recyclable polyester sheath on which a life-size forest is depicted."[24] The unique facade that creates interior spaces with dappled light, also allows birds to see the facade, making it birdsafe (Figure 5.5).

In no small part these designs argue for the inherent biophilic and beneficial qualities of the wood as a building material. There is now considerable evidence about the important aesthetic qualities of wood, and how we seem as a species to be attracted to and affected by contact with wood scent. Biophilic consulting firm Terrapin Bright Green has effectively assembled and summarized the research and literature in their report *The Nature of Wood*.[25] The evidence about wood is clear: we love the way it looks (the warm colors) and feels when we touch it. There is evidence that touching wood panels lowers our blood pressure. And the scent is restorative as well. There are, the way it studies that track eye-movement that suggest that we appreciate the knots and other

FIGURE 5.5   Interface Base Camp, Atlanta, Georgia. Photo Credit: Tim Beatley.

natural elements of wood. I recall my own floors: white ash sourced from a sustainable initiative in Southwest Virginia. We chose a flooring option that fully utilized the qualities of this so-called "character wood": not a square foot of flooring is quite the same and we enjoy seeing the remarkable detail found in its grain and structure. Wood is high in fractals and that might be another explanation for why we love wood so much. And It may be that as the Terrapin Bright Green report suggests there are also significant "associative" or "semantic processing" factors at play: that as we look at wood our brain makes the leap to trees and forests, and as a result "a biophilic response is triggered."[26]

There is accelerating interest in designing larger structures in wood, specifically what has been called mass timber. Here the emphasis is one design and construction utilizing structural elements made of layers of wood glued together. Two techniques in particular are in growing use: one has been called glulam (glue-laminated) wood structures, often used for structural beams and columns, and CLT (cross-laminated timber), typically used for walls. Together these techniques and building products are permitting new, largely timber structures to be designed and built in cities around the world.

There are many advantages to mass timber. Compared to steel and concrete buildings, mass timber structures emit far lower amounts of carbon. Indeed, the wood in the structures itself turns a mass timber building into a form of carbon sequestration. Because of the mass timber elements are prefabricated the time it takes to build a structure can be significantly lower.

One of the best and most natureful early examples of these techniques can be seen in Tye Farrow's Credit Valley hospital and cancer center, near Toronto. As you enter the main lobby of this hospital you find yourself walking into what Architecture Magazine article aptly described as a "grove of timber trees."[27] It is a spectacular example of how to combine living trees in an entrance atrium, with engineered timber columns shaped in a way that makes one feel as if you are walking into and under the canopy of a tall forest.

Farrow has talked about and written about the intent behind the design, noting that in his discussions with patients and families, there was a desire for a design that would give them hope in a time of major worry and uncertainty. They told him they preferred a design that provides a feeling of "aliveness," something that would engender hope.

The Farrow Partners web page provides a bit more detail about the design intentions behind the forested entrance:

> The main design response of the new courtyard lobby consists of four massive structural wood columns, too big to physically wrap your arms around. Each column base is then dividing into three slightly curved columns, which lean outwards, that divide again into numerous smaller individual radial cantilevered beams that support the roof.
>
> The form was intentionally chosen to be reminiscent, familiar, like that of a very old deciduous trees in a meadow clearing; natural, real, authentic, that remained together in the same place year after year after year, in the face of hailstorms and lightning, cold winters and hot dry summers; solid, silent, durable, grounded.
>
> The structure expands out above and over you, creating the perception of a canopy, of shade and protection.[28]

Patients suffering from cancer especially benefit from and need this strong sense of durability, of the strength a forest conveys, and the sense perhaps of entering a space and building that exude certainty and permanence at a time when illness has taken such feelings away. And there is a sense of calmness and protection as one enters this building.

There were a number of technical design issues to work out in this early mass timber structure. One involved the integration of the metal reinforcing elements that would more typically be attached as plates to the exterior of the columns. The concern here was that such exterior plates would interrupt the feeling of trees and forests and so a design for internal reinforcement was devised. Another issue was how to satisfy the fire code. In the end an innovative water misting system was devised.

Milwaukee has become the site of the world's tallest mass timber structure, the twenty-five-story apartment building called the Ascent. Renderings of apartments (there will be the 259) show living spaces with exposed wood beams and wood ceilings. And as with other mass timber projects the construction of this building has substantially reduced labor and transport costs. The structural elements of the building were designed digitally and built offsite. A recent news article about the project describes the assembly process as being not unlike assembling Lego blocks.[29]

There are still reservations about this building technology, but they seem to be diminishing over time. One is about safety and concerns about fire. In the case of Ascent, the structural beams underwent a 6-hour fire test in a US Forest Service testing facility and passed with flying colors, able to maintain their structural integrity. Will residents want to live in timber built high rises? Could this be another obstacle? Ascent developer Tim Gokhman notes that units in the building have been renting quickly and at a higher than market rate, something he attributes to its "wood premium."[30] So perhaps a lack of consumer interest or confidence is not an obstacle either.

Bringing Trees Inside

Another epiphany in the opening weeks of the global pandemic was the worry about the deleterious effects of sitting inside in not very well-ventilated interior spaces. Where possible, we should open windows, we were instructed. But many buildings, certainly many workspaces do not have openable, operable windows. With the growing importance of bio-philic design that is slowly changing. We know that healthier interiors and healthier building design starts with windows: not only as a source of fresh air but also the essential role of natural light, and also ideally home and office views of nature. Some of our best and most impressive recent ex-amples of biophilic design make these qualities paramount.

Trees can serve as a key biophilic design element in efforts to bring more nature inside. The 2021 winner of the Stephen Kellert Biophilic Design Award is a good example of what is possible. The JR Kumamoto Railway Station Building, a twelve-story mixed use structure in Kumamoto City, Japan, has a remarkable interior that includes a thirty foot waterfall and a central core atrium that contains a remarkable amount of nature, including many trees.

More dramatic still, and at a grander scale, is the "Jewel," an interior rainforest located at the Changi airport in Singapore. "Part luxury mall and part indoor rainforest," it was designed by famous architect Moshe Safdie. Opening in 2019, the forest is the centerpiece of a large shopping and entertainment complex. It has become not only a place of respite for travelers moving through the airport (there are some sixty million of those each year) but also a destination for local residents. This domed forest consists of some 900 trees and 60,000 shrubs, planted in tiers, and with a series of walking trails throughout. Most dramatically is a 40-meter-tall circular waterfall in the center, called the Rain Vortex.

On the day I visited a few months before the start of the global pandemic there was a lot of activity in the forest—people lounging around, many strolling and walking, many watching the changing colors projected onto the waterfall. The sounds of the waterfall were prominent and could be heard at some distance away. Despite the many people there it was still a quite relaxing experience overall (Figure 5.6).

New York Times Writer Stephanie Rosenbloom visited the Jewel about the same time that I did and wrote an insightful review, at once not only appreciative and complimentary about the project but also offering a healthy dose of skepticism and critique as well.[31] She describes the feelings of not only being "enchanted" by the

FIGURE 5.6 An interior rainforest and waterfall, The Jewel, at Changi Airport, Singapore. Photo Credit: Tim Beatley.

forest and the waterfall but also observes that there is a fine line "between fantasy and dystopia."

"Looking around, it isn't hard to imagine a future in which everyone lives in domed cities in temperature-controlled, never-ending summers." She also reacts to the corporate feel (naming rights to the forest were sold to a Japanese cosmetics company) and also to the consumptive context of being embedded in what is essentially a large shopping mall (including 280 stores and restaurants). "The result," she says, "is a staggering display of artificiality and nature."

"Something about this might niggle at the back of the mind … Your animal instinct pricks up, and you begin to feel restless, for you know that while these are plants and trees, there's nonetheless a ceiling between you and the sky."

The human-designed nature and artificiality of this domed forest are undeniable of course. There is also no pretense that this is a natural forest. And one could ask how the trees living in this structure are any less natural than say a tree in a planting box along a city street, surrounded by homes and buildings.

In the end, we will want to see such interiors forests not as substitutes for more natural real forests (however we might define that) but they can provide many benefits in cities and they might help in many ways to complement the outdoor forested world, perhaps educating about trees and forests and/or raising awareness but at least providing some positive softening and civilizing of what is often very sterile interior spaces.

There are many obstacles, perceived and actual, to integrating trees and forest into contemporary architecture and urban building projects. They include the technical challenges of making room for trees and designing structural that will support the weight of trees, as well as systems for watering and maintaining trees.

Some of these obstacles are perceptual and aesthetic. In exploring the new and creative ways that green buildings are being designed to include extensive greenery and plants it has become clear that not everyone loves the look and feel of such architecture. WOHA co-founder Richard Hassell told me in an interview that some of his fellow architects have been skeptical about projects like the Oasia Downtown Hotel. "And surprisingly, a lot of reaction initially from architects and people who like architecture was a little bit of revulsion. They didn't like it … Yeah, it seems to offend people who have this idea of architecture as something very pristine and clean and geometric."[32]

It is important to recognize that there are many common times and places where trees, often in pots and containers, make their way inside and help to enhance and soften interior spaces. A large container tree can dramatically change the nature of a room and will not require much expenditure or effort

FIGURE 5.7 A Bonsai Tree in the Collection at the Phipps Conservatory, Pittsburgh, PA. Photo Credit: Tim Beatley.

but will deliver major biophilic benefits. There is, moreover, a long tradition of caring for bonsai trees, in cities around the world, and in many places there are active bonsai clubs where urban residents meet to compare notes and to share their enthusiasm for these trees. And there is some research, fairly convincing, that these small trees and the love we have for them can be highly restorative and therapeutic (Figure 5.7).

Tree Structures of a Different Kind

There is yet another way to begin to think about trees in cities and that is to consider the many emerging "trees of a different kind," that could or do exist there. None of these ideas would be a substitute for abundant real living trees, but they might be supplemental or help to accomplish some of what natural trees provide.

One kind might be human-designed structure that emulates a tree, perhaps blending elements of living nature with manufactured.

An example are the so-called *Supertrees* in Singapore—large metal structures in the shape of trees that provide the biophilic feeling of trees and that also incorporate thousands of living plants. These human-designed and built mega-trees provide many of the functions of trees—shade and habitat, for example.

Another example of a kind of tree of another kind might be the many human structures that serve some of the important functions of trees. We tend not to think of home chimneys in this way, but in many parts of the world they provide an important substitute for cavity nesting sites for species like chimney swifts. Many cavity nesting birds are in decline in part because of the gradual decline in older trees. We need more older trees and we need to manage our urban forest in ways that value the preservation of older trees and makes room for snags and standing dying trees. But chimneys will also be an important habitat and the trend has been in the direction of capping chimneys and installing metal liners that make it impossible for these birds to nest here. There are a variety of creative ideas and initiatives.

Beneficial as well are the erection of swift towers—stand-alone roosting towers often in the parks and nature areas that similarly provide either nesting or roosting spaces for swifts, or both. A recent example is the swift tower built and installed in Piedmont Park in Atlanta.

Cell phone towers disguised as trees is one increasingly abundant kind of alternative tree structure, often raising strong opinions for and against. A result of federal telecommunications law it is now difficult for local governments to stop or prevent the construction of cell towers. The only recourse is to require that steps be taken to reduce their visual impacts and many local governments now require this. As a result, it is not uncommon to see an odd, often too-tall looking, faux pine tree. In Southern California, the towers are designed to look like Eucalyptus trees, pines or even cacti. These disguises are controversial, costing $100,000 or more. In a recent *LA Times* article about the tree towers, a UCLA botanist expresses the sentiment many have about the structures: "When we are losing the distinction of what is 'natural', and what is made by us, that's where it is a little uncomfortable to me."[33]

There are legitimate reasons to object to this trend. Blurring the lines between what is a real or artificial tree is something to be avoided. And the relatively high cost of the fakery is a big concern, with thousands of dollars that could more effectively be steered to planting many new (actual living) trees.

Conclusions

Architecture has been changing in remarkable ways over the last decade or two. Trees and urban forests are playing a more prominent role as a central design feature. We have seen the rise in the forested tower as an architectural style and building category. Trees are being designed into balconies and terraces but are also being brought inside in a major way, as examples of the Jewel in Singapore demonstrate. As designs like the Frick Center in

Pittsburgh or the Manassas Forest school show, building design can connect with the trees and forests around it and can help to educate users and visitors as well. The rapid rise in mass timber buildings, moreover, is a strong nod to the biophilic power of wood, as well as the opportunity to source it more locally and sustainability, directly tying together the growth of a city's-built form to the conservation of forests.

There is no question that trees are adding immeasurably to the quality and enjoyment of these spaces and, in the case of office spaces, to the productivity of the workers who spend time there. We are likely to see even more inclusion of trees in the spaces in and around buildings as we continue to appreciate the benefits of trees and confront the limited spaces in compact cities for trees and tree-planting in more conventional ground level plantings.

There will remain important design opportunities to protect and integrate outside trees and forests into new urban development or redevelopment. This is not an especially new idea, as the story of Frank Lloyd Wright's design for Crystal City tells us. Yet, there remain surprisingly few examples of this today—examples that the real estate and development world badly need. Trees and forests not only provide many ecological and social benefits, of course, but they also add remarkable market value to homes and development projects. We want to live in close proximity to trees and forests even (or especially) in cities, and real estate development to be a force for protection and conservation as much as for cutting trees down.

Notes

1 Lily Cao, "'The Tree That Escaped the Crowded Forest': Lessons From Frank Lloyd Wright's Price Tower," *Arch Daily*, March 25, 2021, found here: https://www.archdaily.com/958989/the-tree-that-escaped-the-crowded-forest-lessons-from-frank-lloyd-wrights-price-tower, accessed June 8, 2023.
2 Neil Levine, *The Urbanism of Frank Lloyd Wright,* Princeton University Press, 2016, p. 226.
3 Wright quoted in Levine, p. 252.
4 Neil Levine, *The Urbanism of Frank Lloyd Wright,* Princeton University Press, 2016.
5 Sandy Deneau Dunham, "5 Architectural Approaches That Are Shaping the Way We Live," *Seattle Times*, September 12, 2018, found here: https://www.seattletimes.com/pacific-nw-magazine/5-architectural-approaches-that-are-shaping-the-way-we-live/
6 Quoted in Dunham, 2018. Ibid.
7 Center for Watershed Protection, "Forest Friendly Development: A Case Study from Oak Terrace Preserve, North Charleston, South Carolina," October 2017, found here: https://owl.cwp.org/mdocs-posts/forest-friendly-development-a-case-study-from-oak-terrace-preserve-north-charleston-south-carolina/
8 Ibid.

9 See Geos.life

10 Ibid.

11 See Discovergeos.com

12 See G. H. Donovan and D.T. Butry, "Trees in the City: Street Trees in Portland, Oregon," *Landscape and Urban Planning*, Vol. 94: 77–83.

13 Stefano Boeri, "Vertical Forest," found here: https://www.stefanoboeriarchitetti.net/en/project/vertical-forest/

14 "Guinigi Tower: Symbol of Prosperity of a Noble Family," found here: https://unusualplaces.org/guinigi-tower-symbol-of-prosperity-of-a-noble-family/

15 Ibid.

16 Quoted in Vladimir Belogolovsky, "Architect Vo Trong Nghia says he loves the idea of living under a tree," Stir World, May 12, 2021, found here: https://www.stirworld.com/think-columns-architect-vo-trong-nghia-says-he-loves-the-idea-of-living-under-a-tree, accessed July 14, 2022.

17 "Amazon shares new details on HQ2 hiring ahead of Career Day 2021," found here: https://www.aboutamazon.com/news/amazon-offices/the-next-chapter-for-hq2-sustainable-buildings-surrounded-by-nature, accessed May 22, 2023.

18 Bullitt Foundation, "Net Positive Energy Over the First Decade," found here: https://bullittcenter.org/2023/04/20/net-positive-energy-over-first-decade/#:~:text=SEATTLE%E2%80%94In%20its%20first%20ten,in%20Seattle%20for%20a%20year., accessed April 28, 2023.

19 See VMDO, "Manassas Park Elementary School + Pre-K," found here: https://www.vmdo.com/manassas-park-elementary-school-and-pre-k.html

20 See "Manassas Park Elementary School," 2011 ASLA Professional Awards, found here: https://www.asla.org/2011awards/456.html

21 "PDX Turns 80," found here: FlyPDX - PDX Turns 80; Shares Early Look at New Airport Design, accessed May 11, 2023.

22 According to writer Benjamin Sutton Steinkamp's tree "proved very therapeutic for Heather Kreinbrink and her husband when their 12-year-old daughter was hospitalized at Cleveland Clinic in 2010. 'It ended up being something we would go to every day for peace and to come to terms with what was happening, Kreinbrink told the *WSJ*." See "More Hospitals Harnessing Healing Power of Art," August 19, 2014, found here: https://news.artnet.com/art-world/more-hospitals-harnessing-healing-power-of-art-81475

23 From a printed sheet displayed on one wall of the structure: "Building for All."

24 Perkins and WIll, "Interface Headquarters," found here: https://perkinswill.com/project/interface-headquarters/, accessed June 6, 2023.

25 William D. Browning, Catherine O. Ryan, & Claire DeMarco, *The Nature of Wood: An Exploration of the Science on Biophilic Responses to Wood*, Terrapin Bright Green, LLC, 2022, found here: http://www.terrapinbrightgreen.com/report/the-nature-of-wood, accessed May 22, 2023.

26 Ibid., p. 10.

27 Logan Ward, "A Tree-Filled Atrium to Inspire Patients," *Architecture Magazine*, July 29, 2014, found here: https://www.architectmagazine.com/technology/architectural-detail/a-tree-filled-atrium-to-inspire-patients_o, accessed May 22, 2023.

28 Farrow Partners, "Credit Valley Hospital," found here: https://farrowpartners.ca/our-projects/credit-valley-phase-one/

29 David Matthews, "Can Money Grow on Trees? World's Tallest Timber Tower Tests Milwaukee Market," *LoopNet*, May 26, 2022, found here: Can Money Grow on Trees? World's Tallest Timber Tower Tests Milwaukee Market | LoopNet.com, accessed, May 15, 2023.

30 Abby Gallum, "In Milwaukee, a 25-story mass timber apartment building makes an Ascent," *Urban Land*, October 18, 2022, found here: https://urbanland.uli.org/public/in-milwaukee-a-25-story-mass-timber-apartment-building-makes-an-ascent/, accessed May 17, 2023.
31 Stephanie Rosenbloom, "My 27-hour Vacation in Singapore's Changi Airport," *New York Times*, December 2, 2019, found here: https://www.nytimes.com/2019/12/02/travel/Singapore-Changi-Airport.html, accessed May 22, 2023.
32 Interview with Richard Hassell of WOHA, 2018.
33 "Trees Give Us Life. These Fake Ones Give Us TikTok on Our Cell Phones," *Los Angeles Times*, June 30, 2021.

6

TREE EQUITY

Toward a Just Urban Canopy

It is perhaps not difficult to appreciate the ways that gray, ugly and nature-less urban streets and neighborhoods depress and discourage their residents. Several years ago I interviewed Kemba Shakur, founder of nonprofit Oakland ReLeaf, an organization dedicated to working with disadvantaged youth in that city. Employed at Soledad State Prison, she observed that to her it was shocking that there were more trees on the grounds of this prison than there were in her Oakland neighborhood. It is hard to overstate the impact of such sterile urban settings on any sense of optimism or hope.

There are significant and longstanding equity issues in urban forestry that have risen to the top of the agenda in a number of cities. This is encouraging, in part a response to the raising awareness in the US about systemic racism, police violence, and especially the killing of George Floyd in the summer of 2020.

In American cities especially, where there is a long history of redlining and spatial segregation, and as a result access to nature, including trees and extent of tree canopy cover, is lower in underserved neighborhoods of color. "Tree Equity" has emerged as a potent call for a more equitable and just distribution of these natural benefits, especially recognizing that neighborhoods with few trees and low-canopy suffer disproportionately from urban heat.

The parallels between historic discriminatory practices like redlining and extent of tree canopy has been the focus of several recent national studies that come to similar conclusions. Dexter H. Locke and his colleagues reach strong conclusions about the lasting impact of redlining on the size of tree canopies in neighborhoods across many cities in the US. They conclude:

DOI: 10.4324/9781003377344-6

Our analysis of 37 metropolitan areas here shows that areas formerly graded D, which were mostly inhabited by racial and ethnic minorities, have on average ~23% tree canopy cover today. Areas formerly graded A, characterized by U.S.-born white populations living in newer housing stock, had nearly twice as much tree canopy (~43%). Results are consistent across small and large metropolitan regions. The ranking system used by Home Owners' Loan Corporation to assess loan risk in the 1930s parallels the rank order of average percent tree canopy cover today.[1]

The inequitable distribution of trees can be seen vividly on the ground in many cities. Washington, DC, represents an example of the kind of income and racial disparities and segregation found in many American cities, further reflected in the extent of trees and tree canopy enjoyed by its residents. Wards 3 and 8 are extreme contrasts. Ward 3, in the city's northwest, is more than 80% white residents, with few families living in poverty and in fact a comparatively high medium family income of \$155,813.[2] The neighborhoods in this ward are leafy and natureful–including affluent addresses that include Woodley Park, Cleveland Park, American University Park, and Forest Hills. Walking around Forest Hills (Figure 6.1), there are

FIGURE 6.1 Trees and Abundant Canopy in Northwest Washington, DC. Photo Credit: Tim Beatley.

abundant trees, and few homes without trees, many quite large and old. It is an aptly named neighborhood. This ward has, according to Casey Trees, the highest tree canopy in the city at 59%.[3]

A few miles away in the city's lower southwest, Ward 8 is quite different. Its population is 91% African American, with a median family income less than a third of that in Ward 3 ($44,665),[4] and with nearly a quarter of its families living below the poverty line. And not surprising, the tree canopy in the neighborhoods in this ward is half of what it is in Ward 3 at a little below 30%. This is not a tree desert, but it is significantly less leafy and less-shaded and further exacerbating the economic and social inequalities in this city.

The good news is that these disparities can be addressed and according to Casey Trees there is considerable potential in both wards to raise the tree canopy levels (to 70% in Ward 3 and 52% in Ward 8 respectively).

In other American cities, where overall canopy is lower than Washington, the differences between neighborhoods are even greater. In Philadelphia, the range is great: canopy in affluent and largely white Chestnut Hill 60% while only 6% in Nicetown-Tioga, where median income is a fourth as high.[5] Chicago, the differences are extreme between the affluent north side of the city, where tree canopy is 60% or higher, and neighborhoods in the south, where canopy is typically less than 10%. This translates into sharp health disparities and also dramatically different life expectancies for residents (shockingly, as much as thirty years).

Tree equity requires cities to take many different kinds of steps to ensure that the benefits from trees and canopy are more fairly distributed. Reforming the plans that cities develop and implement to better integrate and prioritize equity is one important step. Kolosna and Surlock in a study published in Urban Forestry and Urban Greening come to similar conclusions about the inequity in the canopy on two cities they analyze: Chapel Hill and Durham, NC. Utilizing a spatial autoregression model, they conclude that the "presence of non-white populations is negatively correlated with UTC [urban tree canopy] coverage in the 12 study neighborhoods."[6] They looked as well as the comprehensive plans and development management ordinances, concluding there is little mention of trees and or citywide canopy targets.

Especially troubling are the ways that the current tree protection provisions work against addressing existing tree inequity. Tree planting requirements only kick in when there is a development proposal, and so for those neighborhoods where conditions of low-canopy exist, there are few other mechanisms that might result in more trees and canopy. They note that low-levels of existing canopy; low-canopy neighborhoods of color are likely to already be built-out, and so less able to leverage the development approval process to support the planting of more trees. And as trees die over time, the extent of the canopy in these neighborhoods continues to decline, and as the authors note "sometimes becomes the baseline under which new development must comply, compounding the reduction of UTC."[7]

More pernicious is the way in which, in both cities, it is possible for neighborhoods to themselves lobby for a specific, higher tree canopy standard. But they rightly note that is unreasonable to expect historically under resourced neighborhoods to be in a fair position to engage in such lobbying:

"This policy making process allows for resident input into the UTC regulation process, but also provides a mechanism for disparate protection of UTC coverage. Tying ordinance content to the resources available to [the] community and their ability to effectively lobby the municipality to adopt higher standards for their neighborhood can exacerbate disparities observed in historically marginalized communities."[8]

A better approach would be to determine a fair level tree canopy, perhaps a minimum acceptable tree canopy coverage, for all neighborhoods in a city, as an essential matter of fairness and equity. With in turn an emphasis on investments in tree planting in those neighborhoods most deficient as compared to that minimum acceptable standard. And this is precisely what is happening in a number of cities now.

In both Richmond and Norfolk, Virginia, for instance, there are explicit goals to focus future tree planting in underserved neighborhoods, those with canopy under 30%. Richmond, under the leadership of Mayor Levar Stoney, has placed emphasis on expanding parks and nature in historically underserved neighborhoods in the city. Its new comprehensive plan, Richmond 300, sets the minimum target of 30% canopy for all neighborhoods as well as the providing park within a 5–10-minute walk of all Richmonders.[9]

Another positive trend is that equity has also become the central organizing goal in recent urban forest plans. Boston has recently completed a multiyear process of engaging the city in preparing a new Urban Forest Plan, one that has made equity its key focus and priority. More specifically, the Boston Urban Forest Plan states as its first goal, "Equity First.": "Focus investments and improvements in under-canopied, historically excluded and socially vulnerable areas."[10]

The *Philly Tree Plan*, just released in early 2023, also gives priority to equity. Noting that neighborhoods in the city can experience a 22-degree difference in temperature levels, it calls on the city to "prioritize disinvested areas."[11] A priority map guides the plan, developed by overlaying key elements of tree cover, heat exposure, air quality, health risks, income and poverty, and impervious surfaces.

Planting new trees and protecting existing trees in underserved neighborhoods will directly address nature inequities but will also help to address other aspects of spatial inequality. Recent studies show, for example, that pedestrian deaths are much higher in underserved neighborhoods, a result of not only inadequate investment in streetlights and sidewalks and other physical qualities that make neighborhoods safer and more walkable, but also a long history of urban renewal and highway building in and through these neighborhoods.[12] Trees and forests provide not only many essential direct benefits (shading and cooling) but also help to moderate and tame many of the dangers to pedestrians. Investing in trees will importantly not only cool urban neighborhoods and improve air quality there but also slow traffic.

Including explicit equity goals and targets in city forest plans and actions programs is a wonderful and essential first step, but finding the resources to ensure that such plans are effectively implemented is another challenge. Cities should assess each year what has been accomplished and where gaps remain.

There are now additional resources available to cities attempting to tackle tree equity. The Washington DC-based organization American Forests has developed a methodology for calculating a Tree Equity Score for example, that can be used to judge progress. The scores are available for neighborhoods and cities in the US.[13]

A Troubled History of Trees in the US

For many African Americans trees involve troubling histories and induce conflicting feelings. Trees for many African Americans carry a fraught and troubled meaning. Trees for many are symbols of the terror of lynching mobs. Atiya Wells, the founder of a unique urban farm (and forest) called BLISS Meadows, in the Frankford neighborhood of Northeast Baltimore, remembers attending an environmental conference where attendees were asked to conjure up in their minds the memory of a favorite tree from their childhood.

"I remember saying to the organizer. 'You need to rethink your icebreaker.' she told the *Baltimore Sun*. "Privileged people get to climb trees, other people get to hang from trees."[14] Wells made a wonderful presentation about BLISS Meadows to one of my UVA classes in 2019, and she spent much of the beginning explaining the historical context of redlining and segregation in Baltimore.[15]

Many African American writers and commentators note that this anti-nature narrative is too simple and is in many ways wrong. And it ignores the many ways in which African Americans have valued and been actively involved in the protection and stewardship of the natural environment. Carolyn Finney in her excellent book *Black Faces, White Spaces*, makes this point, while also recognizing the ways in which current patterns of racism or racial exclusion reinforce a narrative that people of color are less interested in forests and nature.[16] Her research includes a close examination of the images found in popular magazines, like *Outside*, that feature few black faces on their covers, reinforcing stereotypes and perceptions of nature as spaces of whiteness.

It is a complicated and complex story, notes historian Tiya Miles, and while the legacy of slavery has brutal connections to trees, she notes the need to reject the narrative of "black people being aliens in the outdoors," but "instead recognize the long tradition of African American environmentalism." And while there is a brutal history that plays out in nature, there were ways in which nature became an ally in the struggle for freedom and safety (the shelter provided by "moss-draped forests and rushing rivers").[17]

There are specific inspiring and positive stories of trees, of course. In reading David Blight's biography of Frederick Douglass, *Prophet of Freedom*, there is the story of Douglass's first efforts at preaching, while still an enslaved person working on a farm called Sherwood Forest on the Eastern Shore of Virginia. Here he was allowed to preach on Sundays, to other enslaved men who traveled to hear him. "Under an old live oak on the Eastern Shore on summer Sabbaths, practicing gestures with his arms and shoulders, and modulating the sounds and cadences of his words ... the greatest antislavery orator of the nineteenth century first found his voice."[18]

Still, cities must work not only to ensure there is a fair and equitable distribution of trees and canopy, but they must also strive to fully engage communities of color in the decisions about and care for those trees. Tree planting efforts, for example, cannot simply be organized and implemented in a top-down way but must be planned with and embraced by the neighborhoods where they will need to be appreciated and cared for.

Tree equity requires that the urban forests in cities be welcoming and safe for everyone; not just proximate and physically accessible, but spaces and places where residents feel they belong and can be safe. There has to be a sense of ownership here, a sense that black and brown residents actually belong in such spaces. They must not only feel safe but also comfortable.

Partly this is, again, about a more equitable planning process, one that goes beyond consultation to direct empowerment. But creating spaces where people belong will raise many other questions about perception, control, and feelings of belonging. As Carolyn Finney notes, the diversity of park rangers and urban forest personnel matters. So also does a recognition of the negative presence of police in park and forest spaces, and the chilling effects created by the sight of police uniforms and guns. Policing of forested spaces is a fraught and highly debated topic and one that requires care and listening. Something as simple as changing the uniform styles of urban park and forest personnel helps to send a signal that this is a safe and welcoming space.[19]

How we go about planting and managing the urban forest matters as well and represents the chance to build trust and to devolve some of the direct responsibility. For the urban forest to be experienced and felt as "one's own," it requires cities willing to actively share decisions and to engage in a process that might be described as co-cultivation, something akin to co production or co-planning. Local communities, and neighborhoods of color especially, can be given the power and responsibility to decide where and when to plant trees, with assistance and adequate resources provided.

In what has become a signal study of tree planting in Detroit, Michigan State researchers Christine Carmichael and Maureen McDonough analyzed the neighborhood tree planting efforts of a local nonprofit *The Greening of Detroit* (TGD).[20] More specifically, they sought to explain why some residents indicated they did not want trees planted on their property at all or only under certain conditions. Failure to adequately care for or maintain trees in the past, and a failure to adequately consult with and engage residents, seem to explain much of this. As the researchers conclude:

> Results showed that the non-profit organization exercised sole decision-making power over tree species selection and approaches to tree maintenance, yet expected residents to take an active role in tree care. A decades-long history of negative experiences with trees, lack of city tree care, and inadequate city services broadly caused two-thirds of residents interviewed to either submit a "no-tree request" or only want a tree under certain conditions, including greater assistance with maintenance or the ability to select the type of tree planted.[21]

In some cases, residents were consulted about what tree species to plant, but the overall picture that emerges is of relatively little direct engagement of residents in the planning of these tree-planting events and programs. While this undoubtedly happens often and in a variety of ethnic or demographic neighborhoods, it is a special worry in less affluent and minority-majority neighborhoods that have experienced deep histories of municipal actions (often negative) and programs implemented with little input or direct involvement on the part of these historically discriminated neighborhoods.

Carmichael and McDonough offer a helpful critique about how this well-meaning tree planting initiative unfolded and implications for how to do it differently, and in ways that would better understand the context and history of past treatment. For one, they suggest the need for a deeper and longer process of engagement with residents, certainly beyond the overly short one-month timeframe in which TGD chose planting areas. "Overall, the tree-planting process did not provide a space for discussion or negotiation of species selection or tree maintenance needs and responsibilities," they noted.[22] More opportunities for such discussion is important as well as providing residents some ability to select trees that best suit their needs and preferences.

How a city, then, goes about planting trees and expanding its canopy is extremely important and an essential element of tree equity. As Carmichael and McDonough, greater shared decision making, even when residents choose not to accept a tree, will increase the acceptability and support for tree planting in the community overall.

And the greater the level of neighborhood engagement in the design and planting of these trees, the more likely a resident is to believe that tree is theirs, and the more likely they are to care for it and care about it, and perhaps at some later date stand up for it when it is threatened by some future action. Family and others in the neighborhood, moreover, are more likely as well to see these trees collectively as a community forest, something that is theirs, to which and in which they belong, and in need of collective care and protection.

Carmichael and McDonough also conclude from this research that it is appropriate for cities to expand their metrics of success: going beyond solely looking at not only the number of trees planted (we tend to focus on targets like a million-trees planting goal) or their survival over time, but also how residents are engaged and the level and quality of community engagement. "Instead, to gauge long-term outcomes of tree-planting efforts will involve expanding metrics of success to include community engagement in decision making, resident satisfaction with tree planting, and resident involvement in tree stewardship over time."[23]

Forest Gentrification

Many cities aspire to expand their tree and forest canopies in lower-income neighborhoods and neighborhoods of color, but understandably worry that such green investments will serve to make these neighborhoods more desirable and in turn reduce their affordability and lead to resident displacement. Over the last decade especially there has been a growing awareness of the unintended consequences and strong consensus that cities must do more to avoid or reduce these impacts.

There is little doubt that gentrification has occurred in response to green and biophilic projects in cities in the past, as evidenced by high profile projects such the High Line in New York City. These price and displacement impacts seem clear, though much less clear is what cities ought to or can do about them.

Dan Immergluck in his new book *Red Hot City* documents the pervasive and insidious effects of gentrification and displacement in fast-growing city of Atlanta.[24] Much of this has been stimulated and exacerbated by the Beltline (discussed in a later chapter), a 22-mile linear park and transit project that encircles the city. Despite efforts to utilize the tool of tax increment financing (or what in Atlanta is called a Tax Allocation District, or TAD) to fund affordable housing, relatively little funding has been generated from this and few units of affordable housing have been constructed. The results in this city are devastating for lower-income and residents of color in that city.

Precisely what tools or policies Immergluck feels cities like Atlanta could or should use to better address these patterns of gentrification is unclear but he believes the city missed many opportunities to better integrate and prioritize affordable housing. Redevelopment projects like the Ponce Market, located along the Beltline, were financially underwritten by the public (as public-private partnerships) yet provided scant affordable housing, for instance. "Too often," Immergluck concludes, "policy and planning choices facilitated, fostered, and subsidized development patterns that prioritized the interests of landowners, capital and more affluent households. These choices fostered a heavily racialized exclusion of lower-income families rather than more inclusive development … "[25]

Clearly, Immergluck believes, leaders and project boosters alike have emphasized parks and trails (and trees) over affordable housing. He makes a strong case for why cities like Atlanta need to invest significantly and early in affordable housing.

Some observers conclude that we should be cautious about growing or inserting more nature and trees in cities. There is a position that some hold known as "just green enough," arguing that some attempts at greening or expanding the urban canopy are ok, but that we should be careful not to go too far. We should aspire, they believe, to be just forested enough. While I understand this perspective, I find it at odds with the vision of immersive nature. We want as many trees and forests in cities as we can creatively accommodate. Moreover, the more unusual a low-canopy neighborhood becomes in our cities the less accentuated the price effects will likely be. And a "just forested enough" approach runs the risk of failing to achieve the ultimate fairness and equity we desire in cities. For these reasons we need to look for other approaches.

An alarming and unfortunate approach–not the right one–is a kind of knee-jerk tendency to prioritize housing over trees which in some cities is actually further exacerbating canopy inequality. Two recent examples have made this clear to me, one local (Charlottesville, VA), and one in a more distant, but progressive city (Los Angeles). Locally, the redevelopment and renewal of a public housing project called Friendship Court (and its redesign as a more socially mixed neighborhood) has resulted in the shocking loss of a series of very old trees. Partly the result of an inflexible set of local development rules, the outcome was in part acceptable, I think because of a myopic focus on affordable housing. If the protection of trees at all infringes on the design and planning of such housing it is okay, so the prevailing sentiment goes, to lose them. And there is a corresponding failure to appreciate the benefits of these trees and how their preservation helps to address the fairness and equity of living conditions in the city. The outcome is not a good one–the canopy in this underserved part of town is further diminished, and while residents may get upgraded homes they are left with a hotter, less-shaded, less-livable neighborhood.

A similar dynamic has unfolded in Los Angeles around a proposed affordable housing and transit-oriented development (TOD) called Crenshaw Crossing. Here, a project strongly supported by many for the badly needed affordable housing it will provide in the underserved area of South L.A., apparently disregarded a protected grove of large California Sycamores. Members of the City's Community Forest Advisory Committee appealed the decision and argued strongly for the need to protect the trees. Looking at the pre-development site and the location of the grove it appeared to me that the project could have been designed to save these older trees.

Robin Gilliam, another member of the advisory committee testified before the city's PLUM committee (Planning and Land Use Management) asking as well for a reprieve for the trees. She shared her written testimony with me, which points out the false choice that typically exists between saving trees and providing needed housing.

"Like nearly all community members, I too support increased housing," says Gilliam, who is African American. But she notes the unfortunate attempt by supporters of this project to pit a healthy tree canopy against housing, something she describes as falling "within a long lineage of attempts to divide and distract communities of color from the basic resources and quality of life that are our right."[26]

As a matter of fairness and equity, cities must resist the notion that trees are expendable on the way to expanding affordable housing. Is there an unavoidable conflict between the need to produce more affordable housing in a city and the need to advance more equitable distributions of housing? I think not, but recent examples from both east and west coast US suggest this is a more difficult conundrum than initially thought.

Cities must go well beyond this, of course, and work to address the displacement and unaffordability that results from planting trees and expanding the canopy. There are some promising approaches emerging, however. One can be seen in the example of the 11th Street Bridge Park under development in Washington, DC. Here, through the guidance and support of a community-based nonprofit Hands Across the River, the project has engaged from the beginning the neighborhoods east of the Anacostia River, who will likely bear the brunt of these undesirable effects. Residents have, first of all, been given considerable and early input in the design of the project. Longer term impacts on housing and affordability have been addressed through the preparation of an *Equitable Development Plan,* from which a host of actions have already taken place, from job training (to ensure local residents will benefit from the jobs created by the project), to home buying clubs, to the creation of a community land trust (Frederick Douglass Community Land Trust, to be precise) designed to protect affordable housing (Figure 6.2).

Whether the Washington experiment will prove to be successful is unclear, but its early example and lessons are already being closely followed by planners and urban leaders around the country. This is a positive sign.

It is frequently said (sometimes by this author) that cities, and city planners, need to cultivate a larger and more robust set of tools to address this problem. There is not likely to be one single or primary solution but many. These tools include rent control, community benefit agreements, further and more extensive use of mechanisms such as community land trusts.

Trees and Forests as Engines of Broader Equity in Cities

Trees and tree conservation also represent unusual opportunities to deliver other kinds of economic and social benefits, for instance employment for those unemployed or underemployed in the community, or as a way to ease the reintroduction of formerly incarcerated people back into the community.

New Haven, Connecticut, represents an especially potent example of the latter. Here, a program administered by the Yale School of Forestry's Hixon

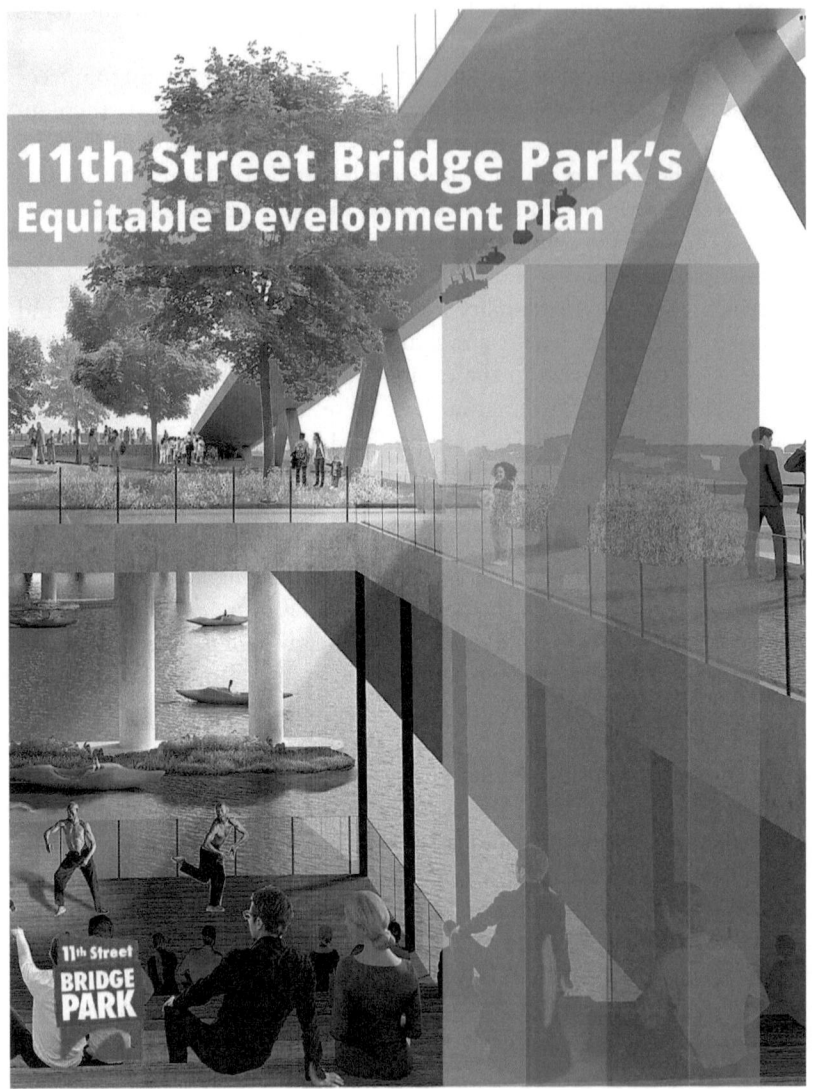

11th Street Bridge Park's
Equitable Development Plan

11th Street
BRIDGE
PARK

FIGURE 6.2 Cover of the 11th Street Bridge Park Equitable Development Plan. Photo Credit: Bridge Over the River.

Center, seeks at once to address not only the need for more urban trees but also provide job training for the formerly incarcerated. An extensive training program is provided, according to Colleen Murphy-Dunning, director of the Hixon Center, along with a variety of so-called "wrap-around services" provided by a local nonprofit called Emerge.

So far, the program has been quite successful, training more than two hundred formerly incarcerated people. The experience for many of them has been transformative. Murphy-Dunny points to how the perception of participants changes, not only by others but also their self-perception changes, from being perceived as a bad actor to one who is doing something good or positive for the community. There is a sense of pride that accompanies the planting of trees in these urban neighborhoods, Murphy-Dunning notes. "Dad planted that tree," a son or daughter might say, and the homeowners who received the trees are grateful, something not commonly experienced by these participants. The relatively low re-incarceration rate shows the power of this program: According to Murphy-Dunning, after two years, this rate is only 15%, quite a lot lower than is usually the case.

Cities can work toward a vision of a more inclusive urban forest in other ways. The variety of potential jobs and employment opportunities connected to planting and caring for urban trees and forests represents a tremendous opportunity to address broader patterns of inequity. Cities can emphasize training programs that provide the skill sets that local residents need.

To what extent is it problematic that most of our urban forestry leaders are white and that most of the tree protection voices and perspectives in cities today are white? Another important strategy has to do with the ethnic makeup of the city staff of front-line departments that deal with tree planting and tree and forest protection. The staff of these agencies ought to look like the city overall and especially the neighborhoods in which they work. This is the idea behind what is often referred to as "representative bureaucracy:" that not only do elected officials need to look like their constituents but also that those who are implementing and enforcing (tree and forest) programs ought to be drawn from and represent the neighborhoods where they and their staff will be working.

The *Philly Tree Plan* makes this point strongly, arguing that "city and nonprofit urban forestry programs must have staff at all levels who reflect the communities their programs seek to serve."[27]

Conclusions

Many cities reflect an uneven and unjust distribution of trees and forests. Forested neighborhoods deliver many physical and mental health benefits, as we have discussed in earlier chapters, and it is profoundly unjust that some residents enjoy those benefits while others do not. In American cities, especially, the distribution of trees and nature follows closely the longstanding patterns of discrimination and spatial segregation, especially historic red-lining maps. Tree equity has emerged as an important goal and a key element of any vision of a canopy city. The main issues of equity are several: a primary one is inherent unfairness of low-canopy environments that are hotter and unhealthy. Many cities, from Richmond, Virginia, to Philadelphia, Pennsylvania, have developed urban forest plans that seek to address these tree and canopy disparities. Cities are increasingly setting minimum canopy targets for all neighborhoods and are focusing on tree planting and protection efforts in neighborhoods with the lowest canopy levels. Ensuring a more equitable process is a key part of the equity agenda; tree planting and protection efforts must engage and directly empower neighborhoods (not simply consult). Connected to this are the many ways in which residents of color especially feel unsafe or as though they do not belong in city parks and forests, things that cities can address in many ways.

Often investments in nature can lead to loss of affordable housing and displacement, something I call here Forest Gentrification. This is a difficult but important challenge for cities to tackle. There are many individual tools and strategies available, but cities must commit to investing in housing affordability at the same time that they seek to plant trees and expand the canopy. Finally, there are many ways, some but not all discussed here, that trees and urban forests can help to address a broad array of larger societal issues around equity and justice, including for instance the reintroduction back into the community of formerly incarcerated individuals. New Haven, Connecticut, has developed one successful model for doing this.

It is important to note that there are many dimensions to the equity challenge facing cities today when it comes to urban trees and forests, and this chapter has just skimmed the surface. There are serious inequalities and inequities in the way urban forests are enjoyed by women, the mentally or physically disabled, and by those in the LGBTQ+ community. Misogyny and pervasive violence against women especially, make a visit to an urban forest a dangerous and unpleasant trip. Tackling these larger issues of gender equity will be difficult for cities but necessary if trees and forests are to be fairly and equally enjoyed by all.

Notes

1 Locke, Dexter et al., 2021. "Residential Housing Segregation and Urban Tree Canopy in 37 US Cities," *npj Urban Sustain* 1, 15 (2021). https://doi.org/10.1038/s42949-021-00022-0

2 DC Health Matters, found here: https://www.dchealthmatters.org/?module=demographicdata&controller=index&action=index&id=131493&sectionId=, accessed June 4, 2022.

3 Casey Trees, 14th Annual Tree Report, 2021, found here: https://caseytreesdc.github.io/treereportcard2021/, accessed June 4, 2022.

4 DC Health Matters, found here: https://www.dchealthmatters.org/?module=demographicdata&controller=index&action=index&id=131493&sectionId=, accessed June 4, 2022.

5　See "Since When Have Trees Existed Only for Rich Americans," *New York Times*, undated, found here: https://www.nytimes.com/interactive/2021/06/30/opinion/environmental-inequity-trees-critical-infrastructure.html, accessed May 18, 2023.

6　Kolosna and Spurlock, *Urban Forestry and Urban Planning*, 40 (2019): 215–223, found here: https://www.ncbi.nlm.nih.gov/pmc/articles/PMC10001874/, accessed May 22, 2023.

7　Ibid, p. 221.

8　Ibid, p. 222.

9　See *Richmond 300: A Guide for Growth*, September 29, 2020, found here: https://www.rva.gov/index.php/planning-development-review/plans

10　See Boston's *Urban Forest Plan*, p. 13, found here: https://www.boston.gov/departments/parks-and-recreation/urban-forest-plan, accessed May 11, 2023.

11　See the Philly Tree Plan, found here: https://www.phila.gov/programs/philly-tree-plan/, accessed May 11, 2023.

12　See Adam Paul Susaneck, "American Road Deaths Show An Alarming Racial Gap," *New York Times*, 2023, found here: https://www.nytimes.com/interactive/2023/04/26/opinion/road-deaths-racial-gap.html#:~:text=It%20found%20that%20Black%20people,as%20that%20for%20white%20cyclists, accessed April 28, 2023.

13　See "Tree Equity Score," here: https://www.americanforests.org/tools-research-reports-and-guides/tree-equity-score/, accessed June 6, 2023.

14　Christine Condon, "At an Urban Farm in Baltimore, Plans for Activism, African American History and Maybe Even Tiny Houses," *Baltimore Sun*, August 26, 2019, found here: https://www.baltimoresun.com/maryland/baltimore-city/bs-md-ci-bliss-meadows-20190826-d47jegag6ze7ff4a6t7m7rilsu-story.html

15　Atiya Wells Presentation to UVA PLAN 1010 class, October, 2019. The entire presentation and discussion with the class can be found here:

16　Carolyn Finney, *Black Faces, White Spaces: Reimagining the Relationship of African Americans to the Great Outdoors*, UNC Press, 2014.

17　Tiya Miles, "Black Bodies, Green Spaces," *New York Times*, June 15, 2019, found here: https://www.nytimes.com/2019/06/15/opinion/sunday/black-bodies-green-spaces.html, accessed June 1, 2023.

18　David W. Blight, *Frederick Douglass: Prophet of Freedom*, Simon and Schuster, 2018, p. 69

19　See for instance the case of changing uniforms of park rangers so they are less intimidating, and other ideas in "Five Ways to Make the Outdoors More Inclusive," The Atlantic, undated found here: https://www.theatlantic.com/sponsored/rei-2018/five-ways-to-make-the-outdoors-more-inclusive/3019/, accessed May 18, 2023.

20　Christine E. Carmichael*, Maureen H. McDonough, "The Trouble with Trees? Social and Political Dynamics of Street Tree-Planting Efforts in Detroit, Michigan, USA," *Urban Forestry & Urban Greening*, Vol. 31, April 2018, pp.221-229.

21　Ibid, p.227.

22　Ibid, p.227.

23　Ibid, p.228.

24　Dan Immergluck, *Red Hot City: Housing, Race and Exclusion in Twenty-First Century Atlanta*, University of California Press, 2022.

25　Ibid, p.221.

26　Written testimony of Robin Gilliam to the LA PLUM, April 4, 2022.

27　*Philly Tree Plan*, Ibid, p.51.

7

IT TAKES A WOODED VILLAGE

One thing is certain, that if trees in cities are to be protected it will require the hands-on involvement of citizens and residents who actually care about those trees. The various local tree protection ordinances discussed earlier are only as strong as their mechanisms for enforcement and while it is tempting to judge enforcement efforts by the technical and legal processes and tools available, it is more meaningful to understand this in terms of support from the public, and really even active participation in enforcement of the law. This is especially true given the inevitable limited resources and staff that cities have available to enforce these laws. It requires many citizens to actively look out for and report unlawful efforts to cut down or damage trees.

We need to reach a point where residents assume a sense of ownership for the trees in their neighborhood. There are programs and initiatives for cultivating this care, but it is true that many residents will already have strong feelings.

Recent examples of residents springing into action on behalf of trees they love and care about can be found in many cities. In Washington, D.C., neighbors witnessed a developer in the process of illegally cutting down several protected trees, including a large heritage tree of more than 100 inches in circumference, and sprang into action. Or the neighbors in the Scarborough neighborhood of Toronto who rang the alarm when trees were being felled there illegally. One of these women, Jean Lu speaks eloquently of the importance of these trees for her. "They are some of the things that give this area a sense of calmness, stability and peace and we have to preserve them."[1]

DOI: 10.4324/9781003377344-7

Among the many recent examples of large old trees being cut down in Washington, is the case of a large red oak cut down in January 2022. A case of a tree protected in theory by DC's Tree Canopy Act, and actually an example of a permit actually denied, yet still cut down. Residents in the Takoma neighborhood did what they could to stop the cut. In this case the efforts were to no end. Yet they still illustrate the important, really essential, role neighbors play.

In this specific case resident Alice Giancola did what she could to ring the alarm and to come to the aid of this tree. "I got dressed hurriedly and I called a couple of neighbors and ran over there … ," she told a reporter for the *DCist*. The news account continues with more of the impressive steps she took that morning: "Giancola confronted the guys who were working on the oak. She dashed off an urgent note on the neighborhood listserv. She called the city's urban forestry division and her local councilman's office. And she called the police."[2]

For her trouble she was yelled out by the property owner. Reportedly, he faces a fine under the law of $72,000, easily absorbed as a regular cost of doing business in the District and seen by many developers as a legitimate option or choice under the law at that time.

The DC police actually arrived on the scene that day but could legally do nothing to stop the cutting. At that point the city had no power to legally stop illegal cuts of heritage trees, though a short few weeks later emergency legislation gave the DC inspectors the legal authority to issue stop work orders.

It is also true that the politics can be shifted in support of tree protection through the work of community groups and nonprofits, working to bring attention to tree and forest conservation issues and actively advocating on their behalf. In Seattle, a resident of the Seward Park neighborhood noticed construction materials being stacked around the base of a large tree, a beloved western red cedar. They alerted the local nonprofit The Last 6000 Campaign, which discovered a permit had been applied for by the developer of the site (who was in the process of replacing a home that had burned down) to remove the tree. Thus began a neighborhood campaign to save the tree, eventually leading to the landowner withdrawing the removal request.

It seemed that many in the neighborhood had grown to love this tree, even naming it "May." One 91-year-old resident, Mordo De Jaen, has lived in the neighborhood since 1959, and was profiled in a newspaper article about the campaign to save the tree.[3] He notes that many of his neighbors love the tree and that it is older than he is.

In St. Petersburg, Florida, neighbors rallied to save a large banyan tree.[4] The tree was a larger-than-life presence on the street, and its loss would have impacted many. In this case, the tree straddled two properties, with one owner wanting to cut it down while the other wanting to save it. Eventually the proponent of removing the tree was convinced to instead trim the tree. One problem is that neighbors do not engage in much thought or direct conversation about the trees around them, and how important they are, in advance of a conflict. Had such a conversation occurred, and had neighbors talked honestly about how important these specific older trees are, perhaps the threat to the tree could have been averted. It not only takes a neighborhood, but it also takes a neighborhood *conversation* to save our trees.

In cities like Atlanta, the nonprofit Trees Atlanta has done much to raise the visibility of trees and to actively work there to protect and preserve the city's vision of itself as a city in the forest. And nonprofit organizations like this often do much of the work of growing and maintaining the urban forest—Trees Atlanta proudly notes that it has been responsible for planting some 150,000 trees in the city since its inception.

The visibility and importance given to trees in the governance structure of the city plays an important role also. Does a city have a tree protection plan of strategy and clear mechanisms for implementing it (and funding it)? Many cities have some form of appointed citizen body such as a tree commission or tree committee, intended to advise city council, and these groups can be highly influential. A case in point can be seen in recent controversy in Augusta, Georgia, over a proposal to cut down trees in its main square, the Augusta Common. This proposal apparently was the result of one commissioner's concern about tree shade killing the grass and also dangerous roots, something that might be the source of lawsuits from visitors tripping on them. In response, the City's Tree Commission held an emergency meeting, pushing back on this proposal and calling for protection of the trees. The chair of the Tree Commission, noting in a news story that there were clearly other options to address the problems other than cutting down the trees.

Replace some sod and fix the sidewalks, said the chair of the Tree Commission, Roy Simkins. But certainly, no need to cut down these twenty-plus-year-old Willow Oaks. One city commissioner coming out of this meeting was convinced by what she heard and by the strong need to protect the trees on the Common.[5] Augusta lacks a fulltime arborist, something surprising to Simkins. Having tree experts on city staff is yet another way to build capacity and give greater attention to trees in the process of city decisions and governance.

We are also in a period of remarkable change when it comes to working and living patterns with potentially significant impacts for cities and for the neighborhoods within them. The global pandemic for many was a time to seek out nature and spaces around their neighborhoods, where the usual indoor office or work environments became less important. It is clear that post-pandemic many workers simply do not want to return to the office, many continue to work from home. Office spaces are experiencing vacancy rates of 50% or higher, signaling the likely permanence of these new kinds of work patterns. Many are now working from cafes and other community spaces at least part of the time, and trees and forests may become coveted outdoor spaces from which to work. These trends also may bode well for the civic life of neighborhoods, as more residents and families are in-situ, and able to enjoy the trees and forests around them, and (hopefully) will have more time to be directly engaged in their conservation and protection.

Strategies for Creative Engagement

As these many examples show, citizen engagement is essential to the protection of urban trees and forests. An important point is to recognize is that for citizens and neighbors to rise to the occasion, to take actions to save specific trees, and to write letters and to stand up for trees in public meetings, and in countless other ways that could help urban trees, requires them to have developed a strong and personal emotional connection to them.[6]

Stephen Kellert has written compellingly of the need to cultivate what he calls a "transformative environmental ethic," of a kind that recognizes and builds upon the deep and lasting affiliation with nature that biophilia supposes.[7] But such an ethic, he believes, will require "developing a deep emotional attachment to and love for nature." As much can be said of trees—emotional attachment and love for trees will represent a necessary precursor to individual and collective action.

How to cultivate these emotional connections and bonds remains an exceedingly important open question, but we live in a time when there are now many creative experiments and initiatives that help to show the way. Each specific city or community will need to develop tools, methods, and ideas of engagement that make sense there, but there are now many examples from around the world to motivate and inspire.

There are many things we can do in cities to highlight the wonderful, beautiful trees that already exist there. For the most part they are an underappreciated backdrop to urban life, performing many important services for us, not the least the provisions of an incredible amount of beauty into our daily lives. Challenging residents to actually pay more attention to these trees is one broad category of approaches to creative engagement.

Yvonne Lynch from Melbourne tells the story of that city's impressive effort at engaging the public in their ambitious urban forest strategy, envisioning a doubling of trees canopy there. The most interesting engagement tool turned out to be email—specifically giving each of the city's 77,000 trees its own unique email address and encouraging residents to send a message to a tree.[8] As she explains, the outcome was surprising as thousands of messages were sent, often touching messages of love.

Lynch evens reports on the interesting result that some trees are emailing each other—one 350-year-old oak tree in Milwaukee emailing an oak tree in Melbourne (with the help of some humans!) "People were really captivated by their ability to connect with trees," she reports. It debunks the idea that technology will always have the effect of distancing us from nature—in this case it has helped to recognize and connect with nearby nature, trees especially.

As another example of harnessing technology, several cities have been using QR codes to connect residents with trees and to help make decisions about urban trees more responsive to citizen opinion. Rotterdam, for example, has recently piloted an initiative of attaching QR codes to many of its trees as part of an effort to better communicate with the public about their management, and in some cases to explain why trees are going to need to be removed, and where and how replacement trees will be planted.

Trees are often the most powerful and obvious way to connect to nature, and there are many wonderful examples of community events and initiatives that center around trees and forests, often as a way to raise larger questions about climate change and global biodiversity loss. One example is currently underway (as I write these words) in the Dutch city of Leeuwarden, in Friesland in the North of the country, where a unique approach is using trees in precisely this way.

Here, though a project called BOSK (*Bosk* is the Friesian word for forest), resulted in 1200 trees become a "walking forest"; trees were moved dramatically through the city, over the course of 100 days. With the help of humans the trees were pushed along in wheeled wagons, on an pre-established route. Hundreds of volunteers assembled each day to push the trees, from 1:00 pm to 5:00 pm, along its 3.5-kilometer course.

The vision is expressed on the BOSK website in this way:

This 'walking' forest gives the trees – and therefore nature – a voice: what can we learn from trees and how does the forest view the human world? A lot is happening in the walking forest, it is always in motion. Residents, companies and (inter)national artists contribute to this. The forest plays with your senses and, thanks to the efforts of hundreds of volunteers, provides a new root network of connection.[9]

This quite dramatic idea comes from a local land artist by the name of Bruno Doedens. I spoke to him by Zoom, about midway through the summer project.[10] He explained a bit of the history and purpose behind the walking trees and some of what has already been accomplished. To him, the walking trees initiative was a way to start to generate a new story. "We need new stories to get better in balance with nature", he said. The current stories—the ones that tell us that humans are at the top of the hierarchy, that say that we are separate from nature—must change. We need to grow up as a species, he told me, pointing out that while humans are a mere 200,000 years old, trees are 370 million years old and as a result have much to teach us.

"What if trees could walk?" Doedens offers as the beginning of an important new story we could begin telling. It is a startling question to ponder, and a most unusual sight to see in action. It is one that has the potential to truly shift our perspective of the world. If this is possible, perhaps anything is possible. And it makes me think of all the new research helping to show that trees and plants are not simply passively stationary but do indeed shift and move and change their immediate and not so immediate environments in many ways we had not previously thought.

It is hard to overstate the logistical challenges of moving 1,200 trees by hand through a city. It takes a lot of coordination, for example with those in charge of roads and traffic (Figures 7.1 and 7.2). Thinking through the

FIGURE 7.1 The Walking Forest, Leeuwarden, Netherlands. Photo credit: Bruno Doedens, BOSK.

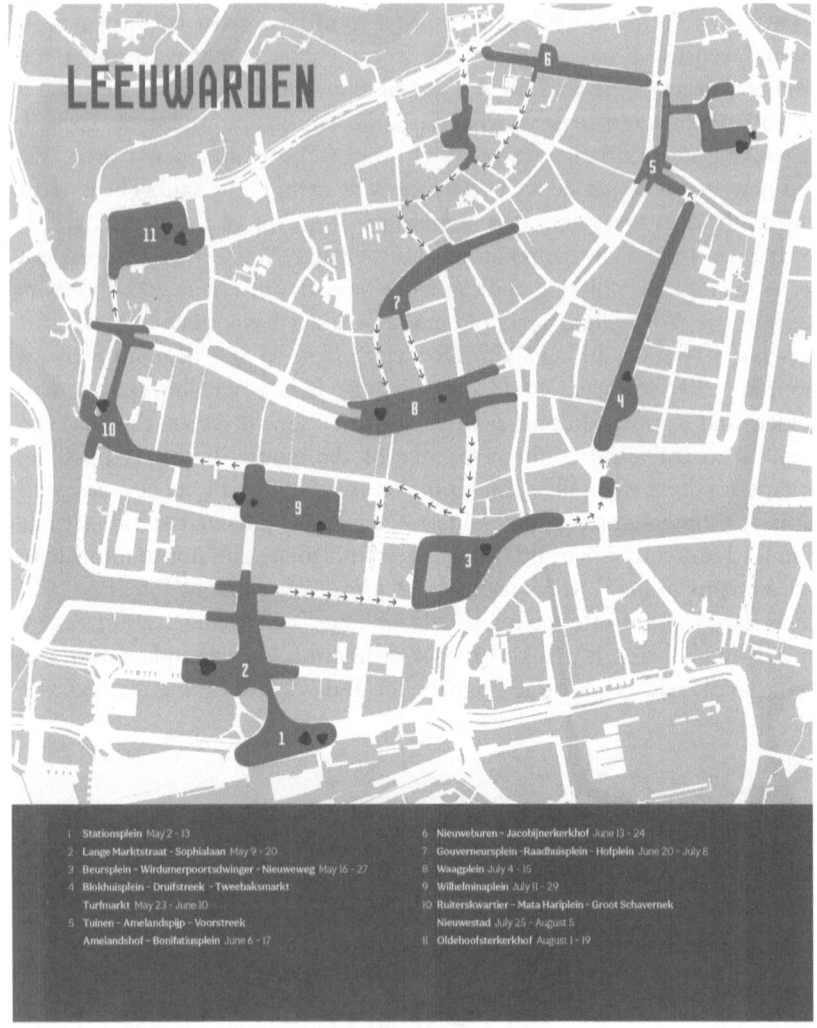

FIGURE 7.2 Map of Leeuwarden, Netherlands. Image credit: Bruno Doedens, BOSK.

containers and the wagons used to move them was also a challenge. Doesden mentions that the trees are quite heavy—some are as tall as seven meters and weigh 1500 kilos each. They needed fourteen specially made hydraulic wagons to push the trees, some wagons with two trees, some holding a single tree. Making sure there were sufficient volunteers to push the trees and also to water them was a coordination challenge also. More than 5000 volunteers were involved in this project, many engaged in actually pushing the trees and the forest along its 3.5 km track around the city. Once the 100 days were over, in August, the trees were then to be planted in pre-identified spaces throughout the city, becoming part of the city's permanent forest.

The trees have touched many different groups and sectors within the city. Schools have been very engaged, for example. At least two school groups joined in the morning each day before the pushing of the trees began.

Especially interesting is how the event helped to stimulate discussion within the religious community. Thirty of the trees were taken into one the city's churches, occupying the space for three or four days, and helping to create at least temporarily what Doedens called a "church for nature." The city's seven churches have gotten together to organize a program of lectures and discussions. This is one of the new stories that Doedens was hoping to help create: how religion must change. "We need an addendum to the Bible," he said, one that sees a more central role for nature.

Partly the immense undertaking shows that anything can be done in cities. And it also changes the sense of what a city is like and what you can find there. As a result of moving the trees through the spaces of the city it has given a very visual and visceral sense of how those spaces could be natureful and forested, and the city could be a profoundly different place.

It has also had the effect, Doedens says, of stimulating many new conversations, and of activating positive change (perhaps we will need to see in a few years the full extent of these effects). Whatever changes happen there in the longer term, there are at least 1,200 new trees in this city.

Not every city will be able to undertake such a grand project. In the end, even with such a heavy emphasis on volunteer labor, it cost several million euros to pull off. They received funding support from the city and province as well as from corporate sponsors. In 2018, Leeuwarden was the official cultural capital of the EU, and Doedens attributes to this much of why Walking Trees was possible. For him the metaphor that makes the most sense is the gradual growth of the web of mycelium that serves as the biological foundation for trees. He is not sure other cities will be able to do what Leeuwarden has done without this social and cultural mycelium.

Bosk was quite a spectacle and quite a media sensation. It has already inspired a number of other cities (Doedens knows of ten cities who have contacted him about doing something similar).

In my own teaching I have found ways to encourage students to pay more attention to the trees and forests around them. Luckily our university is itself a forest, or at least a botanical garden, and there are many incredible trees all around us here. Mostly the students do not see them or acknowledge them. For the students in my Cities + Nature classes I have a specific exercise that asks them to look around, to see the trees on our campus, and choose their favorite. It's part of a larger class requirement of keeping an urban nature journal. Though the students rarely journal to the extent I would like, many take it quite seriously.

I also encourage them to draw this favorite tree and to spend as much time around or under the tree as they can, hopefully returning to the tree occasionally over the course of the academic term (and beyond). While many just scribble a few words and make a quick drawing, or take a quick photo, some of the students spend hours on the assignment. Some of the drawings are quite elaborate, complete with attempts at color (e.g., the vivid yellow of our Ginkgo biloba) and details such as leaf shape and bark textures. I ask them to try to identify the species if they can and many are able to.

The trees they choose are interesting and sometimes unexpected. Several students chose a large southern magnolia, in the courtyard of Thornton Hall, the University's Engineering School. I had not even been aware of the tree, but understand now how it serves to touch and reassure many engineering students who see it and pass by it daily. It caused me to go looking for the tree, to see what I had myself overlooked in my academic neighborhood, so to speak.

Along with their sketches and photos students write about their chosen tree, often explaining why they chose that particular tree and how it made them feel. There is often a memory connected to their choice. One student who had selected a (different) southern magnolia described how seeing this tree took her back to time spent at her grandmother's, where she remembers climbing the tree and collecting the petals of magnolia flowers (then throwing them up in the air, pretending it was a wedding ceremony). It is reassuring that many students have positive memories of actually climbing trees, something I feared had vanished in an era of mostly indoor childhoods.

Support for Neighborhood Engagement

There is a strong argument to be made that if trees are to survive and forests are to grow in cities there must be direct engagement by the residents who live around them. There are many examples of what can be accomplished by just a few dedicated individuals in a neighborhood. Indeed, I think this is one of the most promising approaches for cities to take: to do what they can to enable or inspire, and even financially support, neighborhoods to take direct action to protect and grow the forests around them. This kind of neighborhood forestry, if you will, or neighborhood-based forestry, has a number of advantages.

One example is the story of BLISS Meadows, an initiative to rediscover and care for a small, forested park in Northeast Baltimore Maryland, along with an adjacent vacant lot and empty house. Thanks to the vision and organization skills of nearby resident Atiya Wells, these three properties were brought together to create ten acres of a nature oasis in a neighborhood where there was very little nature before. This mostly forgotten land became BLISS Meadows (BLISS is an acronym standing for "Baltimore Living In Sustainable Simplicity") and has been reimagined as an urban farm and as a place of nature and community gathering space for the neighborhood.

Another compelling neighborhood example can be found a little further south in Washington, DC, where three energetic women in the Brookland neighborhood have taken on the job of regrowing and caring for the small forest at Langdon Park. Located in Ward 5, in Northeast Washington, DC, the park is relatively small—it contains a public pool, a community center and a dog park, and a small two acre forest.

There are many small spaces in neighborhoods in every city that could be transformed from environments that are lifeless and sterile to ones that are nature-infused and forested. So-called forest "patches" can be very small—just a few hundred feet in length and width, and they can be established in the many awkwardly shaped spaces in a city. And together they can improve the quality of a neighborhood and help to build its social capital.

The leaders of the Langdon Park forest patch (Delores Bushong, Mary Pat Rowan, and Allison Clausen) first focused on a small area of the park, about two acres, that was already forested. They started at the edge, in a place where the woods had become overrun with invasives. They worked diligently to cut away these vines, discovering that there were small baby trees underneath. They identified and marked the baby trees with "orange flagging tape," hoping to protect the young trees from being mowed down by the city's park maintenance crews.

Later the team began a more formal research project where they sought to test different planting and management techniques in four different gaps in the forest canopy. The team generated some interesting insights from testing these different "treatments," including the value of mulching as a way to keep ahead of the invasive vines. One startling conclusion of these experiments is just how much regeneration is possible in a small space and how many species (and numbers) of trees can be supported in patches. In the end they counted a total of 706 baby trees, representing a diversity of species, growing in the test plots over a two-year period. Not bad, Clausen notes, in comparison to the roughly 10,500 trees the city has committed to planting throughout the city each year. Tiny Langdon Forest Patch is "holding its weight."

Clausen recently presented their experiences in an online presentation organized by Capital Nature.[11] She defined a "patch" as "a canopy area of at least 10,000 square feet that's at least 100 feet deep." That is small indeed—a little over a fifth of an acre. How important could such a space be? Larger would be better, but even very small patches can themselves be biologically valuable, even hotspots or what EO Wilson has called "micro-wilderness."

A big part of this story, as Clausen tells it, is the value of reaching out and seeking allies, not only in tree conservation groups like Casey Trees, a Washington nonprofit, where there are resources and expertise, but also in local elected officials. The council member for Ward 5, in which the park lies, was invited to the forest patch and actually participated in flagging some baby trees.

The team also thought it would be a good idea to establish the patch more officially as a place—eventually putting up a sign and marking out the boundaries. This, Clausen said, also helped to send the signal that the forest patch was something important. "So we're communicating the value of trees in a very tangible way." She also observes that by attaching a name to the forest patch it becomes "something that you can protect."

One of the key benefits of this kind of work, Clausen says, is the important connection to one's neighborhood, and the chance to do something that can connect them to their neighbors. "We love that we get to meet our neighbors while we're doing this work."

There's an important social dimension to planning and tending patches in this way and there is a credibility that nearby residents have in doing this work that others coming from outside the neighborhood, or from the city's parks department, would not have. "We love that we feel ownership. It's like our space. And that's really motivating. We get more permission to do it because we're neighbors, we are citizens directly connected to this park." They have been leading forest patch walks on Saturdays. The forest patch is now even on Google Maps, making it easier to find.

Whatever very useful insights the Langdon Patch experience has provided about how to effectively grow trees in the city, perhaps the most important lesson is how nearby neighbors, if given the chance and encouraged, can and will take ownership of these spaces. These have built-in incentives and vested interest to be sure, and they do often very much want to see their neighborhoods greener and more natureful.

Claussen suggests that others should follow the example of the Langston Forest Patch tenders, encouraging one to look around and see where their efforts at growing and tending a forest patch could be put to productive use. "There are these little bits of woods all over the city and they could use your help." A close-by forest, one that you can visit often would be best. "Find one that is easy for you to get to, so that you want to go to it often," she says.

In these ways, trees can also help to cultivate community and community bonds. Trees and forests become important spaces for spending time in urban neighborhoods and they help to bring people together and even to heal from trauma and past division.

There is also an opportunity to connect these patches and the agenda of growing and managing neighborhood forest to some of the anchor institutions that typically exist there. These include, for example churches and schools. In Baltimore, Maryland, there have been efforts to establish a network of so-called "resiliency hubs." These are community-based organizations that provide essential services and assistance to underserved neighborhoods mostly within walking distance. It is not only about providing assistance in times of special need, for instance to provide cool space during summer heat waves, but also about providing food and water, and about helping the surrounding residents to more effectively respond to many different kinds of stressors. There are now fifteen- of these resilience hubs in the city, and many have invested (often with city funding) in the installation of photovoltaic panels on their roofs, and rainwater collection tanks. They are also places where food-producing gardens are located.

A ten-acre forest has become a major element of one of these resiliency hubs—the Stillwater Community Fellowship. Here there was a long-forgotten and overgrown forest that has been rediscovered and made a key part of the mission of the church. It has been called PeacePark and seen as a place for healing and contemplation. Recently 1800 saplings were planted to replace the ash trees that were removed because they were infected with emerald ash borers. There is a meditation trail that leads to a stream and pond with meditation stations along the way. There is an apiary and also a community garden at the entrance to the forest. The church's website makes a clear connection between their religious beliefs and the presence of the forest. "We are learning to consider Stillmeadow PeacePark as a clear and singular gift from above," it says. It is a chance to "promote a 'creation-connected focus on spiritual health.'"[12]

The idea and model of resiliency hubs has taken off, and a number of other American cities are pursuing a similar strategy. There remain relatively few examples like the Stillwater Fellowship forest, but the idea of integrating trees and forests makes much sense and as the network of resiliency hubs expands in thar city it is my hope that nearby urban forests and forest patches are included (Table 7.1).

Schools and school grounds represent another potential category of community anchor that could also help grow and protect nearby urban forests. School yards themselves often represent planting opportunities, and there are many good examples of efforts to take up asphalt and

TABLE 7.1 Stillwater Fellowship's PeacePark

"As a church, we have learned to bring what is sacred about the forest into our structured setting. And to take our worship practices out to God's natural creation." "After ignoring our own forest for decades, we became sensitive to our stewardship responsibilities toward God, our community, and the earth. Serving southwest Baltimore with the restoration and development of 10 acres of land; The property of Stillmeadow Community Fellowship. Working with a myriad of Federal, State, City & Community Organizations. In four short years, Stillmeadow PeacePark has risen from the sadness of unhealthy invasive vines and the ash tree epidemic that continues across mid-Atlantic forests."

Source: Stillwater Community Fellowship, undated.[13]

pavement and replace them with native landscaping, vegetable gardens, and also trees. There is a strong argument to be made that every school needs a forest as much as, probably more than it does the conventional play equipment it has (e.g., basketball courts).

Years ago I had the pleasure of visiting an unusual primary school in a suburb of Perth, in Western Australia. Instead of the usual playground it has a native "bush" behind and attached to the school. It was a forested spot of considerable biodiversity and beauty and became a site for doing many things, including a place to work on your homework. There was an afternoon club that focused on the ecology of the site called "bush wardens." I was also impressed by how learning about the bushland found its way into the classroom. Among other things the students in this class learned how to identify the main species of Australian trees found there.

Paying Attention to the Trees around Us

There are now a number of examples of "tree of the year" initiatives that operate in a similar way to raise awareness about the tree species around us. They ask citizens to think about and vote for their favorite trees, or at least their favorites from a list of nominated trees. This can be another strategy for encouraging citizens to think more about and notice some of the most impressive trees around them.

Paying close attention to long-living trees around us can also help to overcome the temporal disconnect between people over time. Bjornerud teaches a class called "History of Earth and Life" that covers, presumably, a quickly-paced primer on the planet's 4.5 billion years of history. I do not recall ever having a class like that but it would be a helpful one to orient any leader about the seriousness of the job they have taken on. Geology, Bjornerud says, is "the closest we may get to time travel."[14] But in cities where there are many older trees nearby, all around, leaders and residents alike are encouraged to deepen their time perspectives.

We have few institutions in which people at all stages of life can gather and experience a unified sense of human community ... We need spaces where, from an early age, children see that they are on an ancient, sacred path that stretches across time, that the richness of life comes from the universal process of unfolding (e-volution), and that growing up and growing old are to be celebrated, not feared. Religious organizations have traditionally filled this role, but we need to be deliberate about finding new venues--choirs, community gardens, cooking schools, oral history projects, bird-watching groups, sturgeon fishing clubs--that can serve as "intergenerational commons."[15]

In San Francisco, born of the pandemic, tree lover Mike Sullivan started organizing self-guided tree walks or walking tours. Author of the popular book *The Trees of San Francisco*, Sullivan had been giving in-person tree tours throughout the city (including of the trees on the rooftop park of the Salesforce Transit Center). In the midst of the pandemic Sullivan heard about an effort in London to organize sidewalk tree tours and thought this might serve as a useful diversion for his fellow neighbors. So he went out one day (as he explains in a recent podcast)[16] with a box of chalk and started writing on the sidewalk information about trees. He would then add an arrow directing residents to the next tree, creating a walking loop.

The tours would then make it onto Sullivan's website "San Francisco Trees," with more information about the trees and photos as well. These clever enticements to walk around the neighborhood and learn about trees were enjoyed by many and suggest the power of low-tech approaches to building awareness about what is around us where we live. Pre- or post-pandemic, walking tours have been an important mainstay of local environmental groups, such as Nature in the City, also in San Francisco. Weekend nature walks and tours of all kinds can also serve to generate employment and income for those who lead them.

But they can also help residents to see and appreciate the many different kinds of trees around them and deepen their knowledge about them, and ultimately foster a sense of caring and concern about them. And it may also help to lay a foundation for neighbors to take action when specific trees are being threatened.

Planting Trees and forests has also become an important way to commemorate people and events in our lives. This practice could be expanded even more broadly and in ways that further solidify the emotional connection to trees and status of trees. Examples abound. In the case of my own family, there is a "Beatley" tree in a prominent park (Fort Ward) in the city in which I grew up. Visiting this tree has been, for my sister, especially a measure of solace. It is not surprising then that we engage in tree planting to remember those who have passed away.

Creative Connections

There are increasingly many creative ways to educate residents about the trees around them, and new technologies that will help foster such connections. The ubiquitous use of cell phones and other digital devices is not only a hindrance and distraction but also a potentially useful tool for educating and building connections. In a number of cities there are new techniques for labeling and identifying trees and helping residents to learn more about particular species and their unique biologies. The use of QR codes, again, is one technique, but utilizing old-fashioned placards in prominent locations in parks, for instance that explain and help to identify common species of trees visitors will likely see can be very effective also.

There are many potential ways to engage artists and the art world in helping make such connections. In the summer of 2019, I paid a visit to an interesting tree project in Hampstead Heath in London to experience firsthand one such example, called the Listening Wood. It is described as "an interactive digital walk around fourteen of the veteran and ancient trees of Hampstead Heath."[17] Aided by a hard-copy map (an online map could also be found), I searched for these special ancient trees. Each of the fourteen had somewhere at their base a wood cut emblazoned with a key word—the object was to locate the word and then text message the word to a site (Figures 7.3 and 7.4). The site would then generate a unique line of poetry involving information about the tree and its particular history.

As I stood beside a pair of ancient English oaks, waiting for a response, a poem appeared in a text message: "These two trees are, to this Holy Shrine, a woodland prayer of Love." A woodland prayer of love was an eloquent way to describe the trees, and the process of searching for and finding the trees and their hidden key words was a fun act of discovery. Poetry is one highly effective and largely underutilized way of connecting to trees. This particular example is described as a collaboration "between artists and technologists" and demonstrates how creative use of technology can foster a sense of connection with trees as well as generate elements of literary delight.

What is the role of imagination? Can poetry help us to see the world around us, including nature and perhaps especially trees, in new ways? Enjoyment of trees and all forms of literature would seem to coalesce, and we should be open to new ways that trees and forests and storytelling in all forms go together.

FIGURE 7.3   Image of the Oaks in Hampstead Park, London, Digital Poetry
Project. Photo credit: Tim Beatley.

Neighborhood Forest Maps: Mental and Magical

The neighborhood scale of a city represents an especially potent place to cultivate new awareness and caring for trees. And there are many strategies for cultivating this neighborhood-level awareness. Learning the names of the species we see around us is one impactful step.

FIGURE 7.4   At the location of these oak trees at Hampstead Heath in London, visitors can find and text a key word (UPSTART in this case), which then generates an original poem about the trees (texted back to the visitor). Photo credit: Tim Beatley.

Recently my wife and I have been noticing the beauty of Loblolly Pines, with their distinctive tall and straight form, and especially their almost reptilian bark pattern. They are a very distinctive tree but yet we were uncertain what species it was. Once we have learned we notice them all over the neighborhood, and that is often what happens with learning species names. Can we truly notice the presence of trees around us without making some attempt at learning at least common names of species?

"Names are passwords to our hearts," says Midwest writer Paul Gruchow.[18] Yet most Americans are not able to identify common species of birds, flowers, and yes (especially?) trees. It may not help that for deciduous trees much of the year there are no leaves to assist identifying a tree. And that even if there is, it is not always easy. Yet we could and should try. If we are to embrace the trees around us as friends and family (and even to treat and name many of them as individuals) then more effort must be made to encourage, help, and facilitate the learning of the names of species, as well as much as we can about their unique biologies. This could happen in many different ways—simple placards are helpful, neighborhood walking tours and self-walking tours are good, as are efforts by friends and family and others who can serve as neighborhood tree docents.

We could also recognize the value and importance of the mental maps we carry with us and to work to modify those mental maps to ensure that trees and nature are more prominent and front of mind. Engaging residents in the drawing of their mental maps is an instructive step. An example of this can be seen in the work of journalist Laura Bliss who, with her colleagues at Citylab, issued a call for readers to share maps of their locked-down lives at the height of the pandemic. The variety of maps she collected, many quite elaborate and beautiful, are a remarkable record of how the outdoors and nature especially played such an important role during the pandemic. "With pens, paper, digital tools, tiles, clay, and whatever else was around," Bliss writes, "nearly five hundred people on six continents sent in breathtaking maps and stories." Sixty-five of these appear in her book *The Quarantine Atlas: Mapping Global Life Under COVID 19.*

The DIY maps that were selected for inclusion in the book are not only beautiful but also an impressive record of the ways that people connected with and sometimes discovered for the first time the nature around them near to where they lived. In perusing the maps the presence of trees in almost all of them is striking. There is the map from a resident of the UK that reports on the discovery of a neighborhood canyon, with an elaborate drawing of what one experiences walking along and through it, including reaching a "dappled wood clearing." There is the resident of Washington, DC, who drew a beautiful map of the trees and water experienced along the Anacostia River. And my favorite, a map by a resident of Sydney, Australia, reporting on the "magnificent trees" on her walk around her block. Her eloquent caption is a fitting companion narrative to the beauty of the trees she drew:

I am dwarfed by enormous gum and fig trees, delighted by butterflies, enchanted by mushrooms in the sidewalk grass. The olive trees hearken to folk tales and distant lands. I am uplifted by the scent of jasmine, alerted by the squawking lorikeets, and beckoned by the rustling bamboo.[19]

The pandemic, as terrible and frightening as it was for many, was also an opportunity to discover, or rediscover, the nature all around us. And we especially seemed to discover and enjoy the urban trees and forests, and these maps show at the least the potential of the CityLab exercise to refigure and recompose (for how long?) our mental maps of our neighborhoods. Such mental and in this actual maps provide the chance to develop a collective understanding of the importance of the trees around us and the chance to collectively draw a line in the sand about what needs to be protected and saved (not only for the pandemic perhaps but also for our everyday, regular lives). For me the role of trees in these maps stands out. And Bliss and her colleagues have discovered a way to make clear and visually compelling the features around them that are most important. Perhaps acknowledging the wonder and respite of those trees nearby will remind one of how much personal and neighborhood value it provides. It might help to cultivate a more effective ethic of caring about that tree and the need to take steps to protect it.

There is much value in working toward a shared neighborhood map that acknowledges the trees and woods that appear in many individual mental maps. It becomes an opportunity to forge a common sense of what makes the neighborhoods special and of the natureful qualities that will help its residents weather the next pandemic or crisis. The map, like all physical maps (though it may be found primarily online in a digital format) serves as a guide to what one can see in the neighborhood, and the chance to layer information and history and biology that might be.

Many cities maintain extensive interactive online tree maps, often organized by neighborhood. These can be very helpful in connecting residents to the trees around them and the kinds of care needed to sustain the neighborhood urban forest. The New York City Tree Map is one of the largest, with more than 800,000 trees recorded on it.[20] The online map allows a user to see and find nearby trees, then to hover over the map point to learn more about the tree (Figure 7.5). Information for the map was itself a grass-roots endeavor, and involved the work of more than 2300 volunteers, coordinated by the nonprofit Trees NY. Information for each tree includes species, trunk size, as well as the last time the tree was inspected by the city. It also allows for information about care needed or given to be recorded for the tree. There is a street view image for each

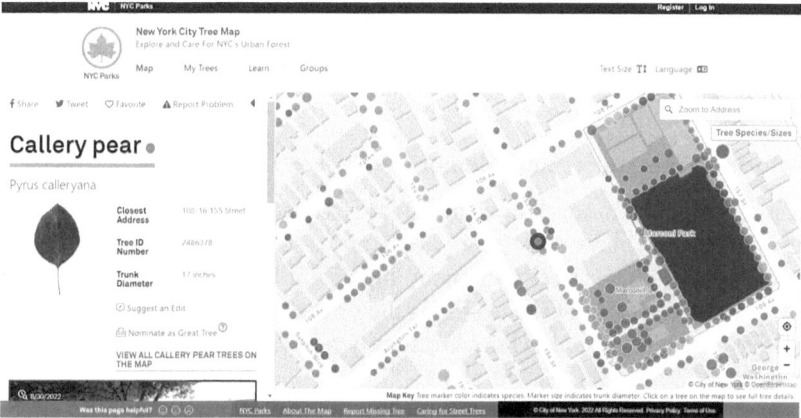

FIGURE 7.5   Screenshot of the New York City Tree Map. Photo credit: City of New York Parks Department.

record, so one is able to see a picture of the actual tree as well as the homes and streets around it. Helpfully the record also calculates the benefits provided by the tree and their economic value. Citizens can also indicate their feelings for specific trees by "favoriting" them—and more than 10,000 trees have been favorited.

I spent a few minutes myself one day surfing around on the map and discovered just how powerful a tool it can be. One can search by species (I wanted to see how many white oak trees there were) or size of the tree which is quite helpful. I zoomed in on trees in the borough of Queens, where the database includes records for an amazing 288,962 trees. I found a 28-inch Catalpa tree on 116th Avenue. The record gave me specific information about what that tree is doing for the neighborhood—it is intercepting 4359 gallons of stormwater, helping to conserve 2048 Kwh of electricity, and reducing carbon emissions by 6132 tons each year. The tree is also removing some 5 pounds of air pollutants annually.

Other cities around the world are also utilizing interactive tree maps. Berlin's tree map helps to coordinate the watering of trees, allowing residents to see which trees around them need water and which do not.[21]

And online maps can also be useful in organizing and coordinating tree conservation work and interest of volunteers in working on behalf of trees. The Philadelphia Horticultural Society (PHS), that runs a very active program of volunteers called Tree Tenders, uses an online map to coordinate the work of these groups (now more than eighty of them in the city). Potential volunteers can find a nearby tree tenders group to join as well as tree planting and other events to participate in. And one of the PHS maps shows tree-planting priorities—neighborhoods and locations where tree planting is most needed, serving as a guide to anyone interested in planting trees.

Capturing Tree Stories and Memories

Trees in cities are more than elements of street furniture but co-actors in the history that plays out there. Residents associate and connect trees, especially larger older trees with many significant life events that occur and play out there. Many of those stories die or disappear when residents die or leave a neighborhood. Many are simply never recorded or captured in any way that could be passed along.

There is much value in attempting to systematically collect the tree memories and stories of important trees. It is both a way better way to not only capture the history of a neighborhood but also can play an important role in strengthening our place commitments. The more we seek to understand and memorialize the presence or the specific trees in a city the more likely we are to care about them and take steps to protect them.

Two professors at Portland State University, Catherine McNeir and Vivek Shandas, initiated a creative project called *Canopy Story* as a way to facilitate the collecting and sharing of memories and history. Specifically, they have created an online map of all trees in the city over fifty feet in height and a webpage where residents can upload stories and images connected to specific trees.[22]

The Canopy Story online map and site are easy to use. One can find a tree and click on the pin to learn more about the tree, including its height, and then to upload a story or memory about the tree. It is also very easy to find and read a story about a tree, with an ability to search by neighborhood. Many of the entries also include photos or drawings of trees.

As an example from the Cully neighborhood, shared by Noelle S:

This grand Oregon oak is an anchor for our community. We celebrated groundbreaking beneath its branches in 2012, hosted our very first monthly work party removing ivy, sheet mulching and planting Willamette Valley natives within its dripline. It has held candle lanterns for musical performances and watched children grow. We are truly grateful to live near this tree.

Or another from Cully, shared by Anna S.:

I often gaze at this beauty as the sun tucks itself over some other horizon for the evening. Holding me anchored, tethered to an earthbound body whilst crows chatter their recounted days before a roost. It feels like an extension of our backyard as it closes the gap between our view south, into the lands of the eternally rested neighbors.

Creating many opportunities in a city for direct experiences with trees and forests—visceral, hands-on contact—is another increasingly important strategy. These include opportunities to visit and walk through forests, to engage in forest bathing, of course, but much more. Touching and hugging trees ought to be encouraged, and there is some evidence that simply doing this delivers considerable benefits. Opportunities to (safely) climb trees ought also to be abundant. Climbable trees in public parks are essential, especially for kids. The chance to take a ropes course and (for older kids and adults) the chance to see what it is like to sit high in a tree and to experience life and the world from that perspective will be important.

Rich Louv, author of *Last Child in the Woods*, has been critical of communities and homeowner associations that have restricted the ability of kids and families to erect tree houses. Many a young person has discovered the joy of being outside, and of being alone with nature, and away from parental oversight, through tree houses. There are some examples of community structures that create public experiences of being high in a tree. Cities such as Perth and Singapore have notable examples of public parks with canopy walks that essentially provide these kinds of tree experiences. In Perth's Kings Park, the canopy walk takes the form of a steel and glass bridge that extends more than half a kilometer in length above the trees (Figure 7.6).

Some years ago, while visiting the Warren National Park in Western Australia I encountered with my family one of several "look out trees." This was a large Karri tree with a series of pegged steps that allowed adventurous visitors to climb to an observation tower 75 meters above (Figure 7.7). Not for the faint of heart, such experiences become lasting memories.

Popup Forests and Other Ideas

Marielle Anzelone is a botanist and conservation biologist based in New York City who has made it her life's work to educate and raise awareness about nature in and around cities. She has received deserved attention around the world for her work pioneering the idea of bringing popup forests to very urban and unexpected parts of cities.

I spoke recently with her about her experiences testing this idea. Her original vision involved the temporary installation of a fairly large temporary

FIGURE 7.6 Canopy Walk at Kings Park, Perth, Australia. Photo credit: Tim Beatley.

forest in Times Square, something that never came to fruition. Instead, she has developed a smaller version, where she transports a small popup forest on an e-bicycle trailer. It was "a forest on the back of a bike," as a story in the *New York Times* called it.[23] The bike trailer was jam-packed with a birch tree and many native plants. She rides her bike from her home in Brooklyn to Times Square, stopping along the way to interact with curious New Yorkers along the way.

Her mobile forest on this day also included a large blowup photograph of a forest (specifically an image of Inwood Hills Park, a remnant ancient forest on the northern end of Manhattan). She was followed by her husband and two small sons, one dressed as a Kerner Blue Butterfly and the other as a white milkweed plant, both species that were once found in New York City.

She described for me some of the interesting reactions of people she encountered along the way. Many were intrigued by what they were seeing, creating opportunities for Anzelone to engage them in conversation about nature and forests. She often asked if they would like to take a selfie photograph, which many did. She described how for some of those New Yorkers she encountered the experience triggered some deeper feelings.

FIGURE 7.7 A Climbing Opportunity in a Look Out tree in the Warren National Park, Western Australia. Photo credit: Tim Beatley.

Part of the effect of this strategy is to see something you don't expect to see in a certain place. It's a mind-shifting effect, a mental stretch. Maybe even a shock to some. But whatever the initial reaction the popup forest does seem to have helped to make trees and forests more visible, and present in an otherwise unlikely, unnatural environment and maybe for a few minutes is able to show that things could be different, that nature and trees could be found in the most unlikely of places. "It helps people feel like there was a tiny piece of Inwood Hill Park plopped into the middle of Times Square, kind of like the wizard of Oz."

"All these stories were welling up in people," she explained as she encountered people all along the way that day. For her she sees especially the need for large cultural change about nature and forests—she is sometimes critical of the often over-emphasis she sees in the reliance on facts and statistics as a way to argue for change. Facts alone, as important as she understands they are, will not be enough to carry the day, and certainly not enough to create the conditions for change she wants to see. For that it will take something else—reaching people at a more emotional level, something her popup forests seems to have done. And it also tends to grab the attention of the local media, helpful in reaching a larger audience.

Anzelone does other things these days to educate New Yorkers, including giving free plant walks.

Mystery and Wonder Could Be Front and Center

Part of what will be necessary is to strengthen a sense of the mystery and magic of the trees and forests around us. And to understand them as essential habitats for many different creatures including humans.

The more we learn about the complex biology of trees and the more remarkable they appear to us. Several years ago, I had the opportunity to attend a "medicine walk" through High Park in Toronto, with noted Canadian-Irish ethnobotanist Diana Beresford-Kroeger (Figure 7.8). It was an eye-opening experience not only in terms of the content and knowledge she shared that day but also the unique ways she helped the group connect with and see the trees around us as we walked. She would stop at specific trees, explain their deeper history, occasionally hugging or embracing a tree and expressing her clear affection for them (as my photo earlier, taken that day, clearly shows). Later I read with great enjoyment, her memoir *To Speak for the Trees*, and watched her film Call of the Forest (later screening this in several of my classes).

In an interview in *The Tyree*, Andrew Nikiforuk notes Beresford-Krueger's immense curiosity and knowledge:

> She can tell you that the Laurel tree has always served as the best defense against tsunamis in Japan because their roots travel 30 feet down into the ground. The balsam fir, which the Cree called piko-wahtik, releases anti-bacterial compounds in its aerosols. The green needles of the white pine (Pinus strobus) make an excellent treatment

FIGURE 7.8 Diana Beresford-Kroeger on a Medicine Walk in High Park, Toronto. Photo credit: Tim Beatley.

for bereavement and sadness. It, too, serves as hospital for boreal song birds. The kingnut hickory (Carya laciniosa), her all-time favourite tree, was used by Indigenous people as a source of primary protein. They regarded it as an anti-famine tree. And on she goes.[24]

Indigenous knowledge these authors declare is an important counter-point to traditional western science. Sometimes referred to as "two-eyed seeing," there is a growing recognition of the importance of both in understanding the biology of trees and forests and in their management.

We will learn more about trees that will continue to deepen our appreciation and amazement of them, but there may always be many things about them that we do not know and may never know. This is part of their mystery, I believe.

UBC forest ecologist Suzanne Simard's work further expands our sense of the remarkable, showing how trees are bound together by their mycorrhizal networks and can share nutrients, carbon and information. She has cleverly called it the "wood wide web."

Trees are "in a web of interdependence," she says, "linked by a system of underground channels, where they perceive and connect and relate with an ancient intimacy and wisdom that can no longer be denied."[25]

She feels a deep kinship with trees. Recognizing her own personal family history of logging she has a sense of duty about the work she is doing: "With taking something comes the obligation to to give back."[26] The sense of kinship she believes we need to cultivate, and the sense of trees as sentient beings, draws a close connection to first nation views. Trees are seen as people.

Simard tells the First nation story of the Subiyay:

This begins by recognizing that trees and plants have agency. They perceive, relate, and communicate; they exercise various behaviors. They cooperate, make decisions, learn and remember--qualities we normally ascribe to sentience, wisdom, intelligence. By noting how trees, animals, and even fungi--any and all nonhuman species--have this agency, we can acknowledge that they deserve as much regard as we accord ourselves.[27]

Many plant ecologists point out the "plant blindness" that has tended to prevail in our society. Animals are animated and interesting and as such we tend to pay more attention to them: plants including trees are viewed as passive and stationary and as such perhaps less interesting or wondrous. Yet that is changing in response to the research of Simard and others.

Simard has written about what she calls a "superiority complex" in thinking about plants. We tend not to see plants as "highly evolved creatures, but rather as inanimate, passive, and inferior species. We have constructed a simple vision of plants as lacking in intelligence, agency, or sentience. We have relegated them to the lowest rung of a hierarchy that is headed by humans."[28]

Research suggests that plants and trees are not passive or stationary at all—one recent study shows how trees move and shift over the course of a day, largely in response to shifts in the hydraulics of water. Some have described this movement as akin to a tree's heartbeat, though of course trees do not have a blood-circulating heart in the same way a mammal does. Australia tree researcher Sebastian Pfautsch talks of plants having a "pulse" as water moves through and around trees over the course of the day.[29]

There is recent evidence that plants have a voice, in the sense that they produce a variety of sounds. A recent study published in the journal *Cell* reports on the sounds that emanate from plants when they are under stress. Likely the result of a process called cavitation,[30] researchers discovered sounds in an ultrasonic range that could likely be heard by other nearby organisms. It is an intriguing idea that plants are being heard and reacted to by other nearby organisms and perhaps other plants, and that listening for these sounds could be a useful way of detecting stresses that urban managers and others might be able to respond to (watering parched trees that might be on the verge of dying). The authors speculate also about how such plant voices could lead to responses by other plants and trees. "Could plants potentially respond adaptively to the sounds of their drought-stressed or injured neighbors?"[31] the authors wonder. Perhaps it is possible that trees and plants more generally could develop a variety of helpful responses to these sorts of plant distress calls? Suzanne Simard's research has already shown that mother trees share nutrients and carbon and are capable of sending warnings about pests and other potential threats to trees. Together this emerging research deepens our understanding of trees and our sense of how much more complex plant life may be than we thought.

The work of plant behaviorists, including Monica Gagliano, has shown through clever experiments (many described in her book *Thus spoke the Plant*) that plants are able to remember and to learn, and in one experiment to follow the sound of water (she uses the word "choice" in describing this behavior) the sound of water.[32] Trees no doubt have similar abilities and the more we learn about how plants engage in complex movements and animated action (though perhaps less perceptible and noticeable than we humans are practiced at discerning) the more likely we begin to see them as intelligent and not so different from humans perhaps.

Trees die of many different causes, but they are unlike humans (and other mammals) in the sense that they do not age at the cellular level. Humans experience what is known as "senescence" or the "time-related deterioration of the physiological functioning necessary for survival and fertility."[33] Trees do not experience this deterioration and at least in theory "could live indefinitely,"[34] as a recent Scientific American article declares. Trees are killed by many things (by pests, lighting strikes, weather) but they do not die naturally, leaving some to conclude they could live forever.[35] And no wonder then that we have trees such as bristlecone pines and sequoia trees that are able to live for thousands of years. This knowledge contributes to the uniqueness and wonder of trees but also provides practical knowledge that in the absence of a deadly event older trees, even in stressful urban settings, will not necessarily succumb to some preconceived life-expectancy but could grow yet older proving for many additional years the benefits they deliver to urban residents.

The biology and morphology of trees lends itself to mystery. Much of the life of trees and forests of course happens out of sight and underground. We could do a better job helping to visualize this hidden arboreal nature. Some have tried to do this by excavating and drawing the root systems of different species. The Wageningen University roots systems collection is a wonderful example of how to re-envision the fullness and full biology of trees. There are, here, more than 1000 of these drawings, largely produced by two Austrian botanists, who have been pioneers in the somewhat obscure and undervalued subject of "root ecology."[36] The comparison of root structure of different tree species is particularly fascinating, showing how for some species the roots head downward and deep, while others exhibit more lateral spreading of their roots.

These wonderful root drawings in turn raise questions about how we typically display or depict trees in cities. Perhaps understandably we seem only to pay attention to those above ground parts of the tree, its trunk and boughs and leaves, forgetting the magical world of underground roots and fungi and the complex nutrient and water flows that underpin the survival and thriving of the trees around us.

There is so much that is surprising and wondrous about trees and forests. There is no lack of insights and science to activate this wonder, but the challenge becomes how best to deliver this information and to effectively share it beyond those who are already committed tree enthusiasts and lovers. Paying more attention, again, to the trees and forest around us in our neighborhoods is a bit part of the answer. Making it fun and meaningful through, for example, urban tree walks and forays, and urban tree expeditions that might captivate especially young kids and families.

How much more reverence and respect for the trees and forests around us might be brought about if we had more tree walks like these, and more people like Beresford-Kroeger discussing the remarkable and unique species-specific natural histories of trees. Neighborhood tree walks of this kind ought to be a common event, something every child and adult has the chance to experience and perhaps lead. They will not cost much, and do not depend on large infrastructure investments or parks department budgets.

Some form of tree walk can be found in many other cities, and in fact the global COVID pandemic has helped to spawn some of them. As unique and different as Beresford-Kroeger is, there will be individuals who are knowledgeable and passionate about trees in every city, probably in every neighborhood.

Could we begin to see the trees around us where we live, the leafy neighborhoods of cities, as larger communities of life. We are not separate from the forest, even as residents of a big city, but a part of the forest, part of nature. This is the essential insight of environmental philosophers like Arne Naess, who coined the term *deep ecology* to refer to the possibilities of a larger, more transformative sense of connectedness, a sense of being at one with a larger whole.

If we pay sufficient attention to the nature around us we will tend to appreciate this magic. The trees and urban forests are sufficed with non-human living creatures that can provide delight and pleasure even in the dark of winter. While writing this book, I had such an experience walking in the urban forest that is my neighborhood. We were treated on that gray winter night to a magical sound show, in the form of a pair of great horned owls engaging in an hour-long "duet." One would call out, with a response after a few seconds from the others. I was not even aware such an exchange was possible or common. It was also a visceral and memorable demonstration of what other magical, wondrous life occupied the branches and the world above and around us. Without the owl serenade that evening we would not have known or even imagined the presence of these majestic birds.

The soundscape that emanates from trees is itself a magical concert. Without trees so many of the sounds of nature would be absent from cities—cicadas during the day, tree frogs, crickets and katydids that fill the senses in summer (at least in eastern North America). Even the very sounds of the moving leafs and branches; the windblown flutters, the creaking of boughs, all a part of what sound artist Bernie Krause calls geophonies (as contrasted with biophonies).

In his book *The Song of Trees*, David George Haskell describes in great detail the many unique and distinctive sounds different species of trees make. Their songs are all different in this way—the sounds of rain drops as they fall on leaves and branches of a distinct size. He describes in an opening chapter the sounds or songs of the Ceibo tree of Ecuador, in particular the sound of rain as it falls and hits the various parts of the tree from crown to leaf litter. "Every falling water drop is a tap against leafy drum skins. Botanical diversity is sonified, calling out under the drummer's beat."[37] And then there are the many songs and voices of the animal life inhabiting the layers of these trees from birds to frogs to insects. Each tree will have its unique sonic signature it seems, and we should listen very closely to fully understand and appreciate the life of a tree, to hear its own special "voice."

Recently, sound recordings have been made of the large, interconnected forest of Quaking Aspens in southern Utah's Pando. Pando (Latin for "I spread") is a clone forest; really a single living organism, and as a result can be considered the world's largest tree by weight or size. Sound artist Jeff Rice and Friends of Pando director Lance Oditt teamed up to install a series of microphones, some embedded deep in the soil, to record the sounds of this remarkable living organism. What results is an intriguing and unusual sound map of an above ground (and below ground) forest.[38]

Listening carefully to the sounds of trees, not only the obvious sounds of rustling leaves and bending boughs but also the more hidden and deeper sounds of vibrating roots, is another way of enjoying the mystery of trees. And another way of hearing the distinctive voices of the living trees and forests around us.

Conclusions

"It takes a wooded village" is a key message of this book. Adopting tree planting targets or a strong tree protection code are all important steps, but implementing and enforcing those provisions will require the active engagement and participation of citizens. A code is only as strong as the enforcement measures and these are only as effective as the eyes and ears of citizens to alert public officials to when the codes are being ignored and permits eschewed.

It may be at the neighborhood level where the greatest potential exists to cultivate a citizenry that cares about and stands up for trees and forests. It is here where a threatened tree is likely to be learned about, where some form of activism or intervention might make a difference. And it is here where many of the most significant and meaningful activities can occur—tree walks can be organized, trees in backyards and vacant lots can be planted and cared for, where the many benefits of trees may be most apparent and appreciated. Many of these citizen-based and community-based activities also serve to build a village, in the sense that they are investments in stronger communities and neighborhoods that will pay larger dividends in the longer run. There is evidence that engaging in things like planting trees can serve for many as a gateway to other civic activities and engagement.

Notes

1 Quoted in *CBC News*, "City Investigating After Oak Tree Allegedly Cut Down Illegally At Scarborough Home," December 21, 2021, found here: https://www.cbc.ca/news/canada/toronto/city-investigating-tree-removal-scarborough-1.6292541, accessed May 17, 2022.

2 Jacob Fenston, "Developer in Takoma Cuts Protected Heritage Tree, Over Protests From Neighbors," *DCist*, February 3, 2022, found here: https://dcist. com/story/22/02/03/developer-cuts-heritage-tree-takoma/, accessed May 18, 2022.

3 Aguero Pacheco Flores, "Seward Park Neighbors Come Together to Save an 'Exceptional' Tree," *South Seattle Emerald*, March 10, 2022, found here: https://southseattleemerald.com/2022/03/10/seward-park-neighbors-come-together-to-save-an-exceptional-tree/, accessed July 28, 2022.

4 "A St Petersburg Neighborhood Banded Together to Save Their Banyan Tree" *Creative Loafing*, April 22, 2020, found here: https://www.cltampa.com/ tampa/a-st-petersburg-neighborhood-banded-together-to-save-this-banyan-tree/Slideshow/12387806, accessed May 5, 2023.

5 George Escola, "Removing Trees From Augusta Common Slammed by Tree Commission," *WJBF News*, March 18, 2022, found here: https://www.wjbf. com/news/removing-trees-from-augusta-common-slammed-by-tree-commission/, accessed May 24, 2022.

6 Stephen Kellert, Birthright: People and Nature in the Modern World, Yale University Press, 2014.

7 Stephen Kellert, *Birthright: People and Nature in the Modern World*, Yale University Press, 2014, p.196.

8 Yvonne Lynch "Enhancing Urban Ecology for the City of Melbourne," webinar presentation to the Biophilic Cities Network, found here: https://www. youtube.com/watch?v=W6tpNXXUmow, accessed July 12, 2022.

9 "Bosk," found here: https://arcadia.frl/projecten/bosk/, accessed July 13, 2022.

10 Interview with Bruno Doedens, July 21, 2022.

11 Presentation by Allison Clausen about the Langdon Park Forest Patch, to Capital Nature, via Zoom, November 17, 2022.

12 See "PeacePark," found here: https://stillmeadow.community/peaceparkproject, accessed May 22, 2023.

13 "Nature Peace in Baltimore," found here: https://www.stillmeadowpeacepark. com/, accessed May 22, 2023.

14 Bjornerud, p.178

15 Bjornerud, 176

16 "Total SF" podcast, found here: https://www.sfchronicle.com/podcasts/article/ Listen-Branching-out-with-S-F-s-tree-maestro-17540051.php, accessed April 20, 2023.

17 "The Internet of Trees—The Listening Wood," found here: https://connected-environments.org/portfolio/the-internet-of-trees-the-listening-wood/

18 Paul Gruchow.

19 Ibid, p.115

20 You can find the New York City Tree Map here: https://tree-map. nycgovparks.org/

21 Susana F. Molina, "'Gieß Den Kiez': Giving Trees in Berlin a Chance to Outlive Their Planters," *The Urban Activist*, August 31, 2021, found here: https://theurbanactivist.com/idea/gies-den-kiez-giving-trees-in-berlin-a-chance-to-outlive-their-planters/, accessed June 1, 2023.

22 See https://canopystory.org, accessed June 1, 2023.

23 See "A Forest on the Back of a Bike," *New York Times*, May 11, 2018, found here: https://www.nytimes.com/2018/05/11/nyregion/new-york-today-a-forest-on-the-back-of-a-bike.html, accessed April 18, 2023.

24 Andrew Nikiforuk, "Talking with the Botanist Who Talks to Trees," *The Tyee*,

February 24, 2020, found here: https://thetyee.ca/News/2020/02/24/Talking-With-Botanist-Talks-To-Trees/

25 Suzanne Simard, p.4.

26 Ibid., p.293.

27 Ibid., p.294.

28 Suzanne Simard "Forward" in Monica Gagliano, *Thus Spoke the Plant*, North Atlantic Books, p.x.

29 "Measuring the pulse of trees," March 16, 2015, found here: https://phys.org/news/2015-03-pulse-trees, accessed April 11, 2023.

30 Described as "a process where air bubbles form, expand and collapse in the xylem, causing vibrations," in Itzhak Khait et al., "Sounds Emitted by Plants Under Stress Are Airborne and Informative," *Cell*, 186, pp. 1328–1336.

31 Ibid.

32 See Monica Gagliano, *Thus Spoke the Plant: A Remarkable Journey of Groundbreaking Scientific Discoveries and Personal Encounters with Plants*, North Atlantic Books, 2018.

33 For more detail on the biological processes of aging, see National Institute of Health, "Aging: the Biology of Senescence," found here: https://www.ncbi.nlm.nih.gov/books/NBK10041/#:~:text=Entropy%20always%20wins.,synthesis%2C%20and%20the%20organism%20ages, accessed April 11, 2023.

34 Robin Lloyd, "Trees Have the Potential to Live Indefinitely," Scientific American, December 24, 2021, found here: https://www.scientificamerican.com/article/trees-have-the-potential-to-live-indefinitely/

35 Robin Lloyd, "Trees Have the Potential to Live Indefinitely," *Scientific American*, December 24, 2021, found here: https://www.scientificamerican.com/article/trees-have-the-potential-to-live-indefinitely/, accessed April 11, 2023.

36 They are Lore Kutschere and Erwin Lichtenegger. See the Wageningen root system collection, here: https://www.wur.nl/en/newsarticle/drawings-of-forest-trees-in-the-root-systems-collection.htm, accessed May 22, 2023.

37 David George Haskell, *The Song of Trees: Stories From Nature's Great Connectors*, Viking, 2017, p.6.

38 E.g. see Jennifer Nalewicki,"Listen to the sounds of Pando, the largest living tree in the world," *Live Science*, June 1, 2023, found here: https://www.livescience.com/planet-earth/plants/listen-to-the-sounds-of-pando-the-largest-living-tree-in-the-world, accessed June 6, 2023.

8
TREES NOT CARS

"I'd Rather Be a Forest Than a Street," as the famous lyrics from a Simon and Garfunkel song go.[1] Most of us would probably share this sentiment, and we would rather have more trees in our urban neighborhoods than cars. Already too much priority is given to cars and car mobility often with negative consequences for health and quality of life—more cars means less walkability and more danger for pedestrians, worse air quality, and often excessive noise, among other ills.

But perhaps by rethinking our automobility and working to reduce our dependence on cars we can also create the opportunity to shift our cities and urban neighborhoods in the direction of a more forested urbanism?

This is clearly a big part of the answer to how and where cities can fit more trees. For one, a remarkable percentage of the space of a typical American city is devoted to cars in one way or another, especially in the form of parking: often 25 to 40% of an American city's space is taken up with cars. Parking guru Donald Shoup, author of *The High Cost of Free Parking*, has estimated that in the US we might have as many as two billion parking spaces, and as many as eight spaces for each car on the road.[2]

Rethinking Parking

One important category of ideas is to find ways to not only reduce the numbers of parking spaces we require but also to reimagine what that urban land could look like, and especially pursue the chance to afforest or reforest these spaces.

DOI: 10.4324/9781003377344-8

Ironically the very word "parking" has an earlier meaning that has been largely forgotten. It turns out the original use of the term "parking" was to connote places for planting trees and greenery. This is the interesting idea, and a largely forgotten history, that dates to the 19th century in Washington, DC, where, before the appearance of automobiles, "parking" referred to the greenery and natural spaces between the curb of a street and a person's property line.

In L'Enfant's Plan, Washington was designed as a city of dramatic vistas and wide boulevards. In the mid- to late-1800s, these very wide streets became for some a liability, and a source of irritating dust. The idea was proposed to narrow the streets creating large adjacent spaces for parklands and greenery. Through the Parking Act of 1870 Congress created a new framework putting this idea into practice.

According to the DC Office of Planning, the Act "designated part of the right-of-way immediately next to private property as a park area to be maintained by the adjacent property owner. This area was to be landscaped and is still referred to as 'parking', a term that predates the emergence of the automobile as a dominant part of American culture."[3] As the car emerged in the early 1900s in a big way, and really began to dominate streets in cities, the meaning of the term "parking" changed markedly and took on the more typical meaning today.

It is possible of course to forest-up the parking we have and many cities have adopted landscaping standards that seek to ensure that after a certain number of years a parking lot will be mostly shaded. European cities commonly require a shade tree every three to five car parking spaces. These are still largely spaces for cars but there are benefits, of course. One is that these spaces are dramatically cooler and more pleasant. Without the trees many larger parking areas are dangerously hot and can be tolerated only through the constant running of air conditioning units in cars (Figure 8.1).

From Parking to Forests

A positive trend is in the direction of questioning the need to require new development in cities to provide parking. This makes sense for many reasons, not the least of which it is very expensive and adds greatly to the costs of housing. Buffalo became the first US city to scrap the minimum parking mandate, while Austin, Texas, is one of the most recent cities to follow suit. This step will not help much with all the parking already in place but it will hopefully re-balance the space demands and allocation in cities in favor of nature in trees.

But better still would be to depave and to take out as much existing parking as possible replacing it with trees and greenery and wild nature. There

FIGURE 8.1   Trees Shading a Parking Lot in Charlottesville, Virginia. Photo Credit: Tim Beatley.

will be opportunities in every city to repurpose existing parking, and this is already happening in some places. There are wonderful emerging examples of efforts to make this shift.

Richmond, Virginia, is a city attempting to do this as a way to expand its tree canopy and as a response to tree inequity in that city. In many parts of this city there are places where expansive parking lots have created tree deserts.

Converting at least a portion of surface parking to forested land is a potentially effective strategy in many other cities. An interesting project there has been underway to demonstrate what is possible. Called "Parking to Park," it envisions the transformation of a several-acre surface parking lot at the Richmond Natural History Museum into a new park and forest. Here parking spaces will be replaced with a (six-acre) public park, boasting many new trees. "ADA-accessible paths and casual seating areas, shaded by overstory and understory trees, provide opportunities for guests to gather, exercise, reflect and connect."[4] The park, called "The Green," "will serve as both a welcoming park and a living classroom."[5]

The plan is to replace the 381 spaces with spaces provided in a new three-story parking deck: "The long-term hope is that other property owners and state institutions in the area with large surface lots may see the value of depaving and similarly shift their parking underground, stack it in a deck, or get rid of it altogether as Broad Street transitions to a more transit-oriented area."[6] That seems more and more likely, as transit improves in the city, including the relatively new Pulse BRT. To encourage patrons to get there in some way other than by car, the museum offers free admission to anyone arriving by transit. The location is central, adjacent to Broad Street, and what the museum calls its "front yard."

One hundred percent of the trees to be planted in the park will be native to Virginia, with 70% found in the Richmond region. Landscape architects Glave and Holmes were hired to develop the planting and landscape plan.

In the Netherlands, the increase in interest in tiny forests represents an opportunity to convert some car spaces to trees. As mentioned earlier in Chapter 4, projects like the tiny forest at Muziekplein in Utrecht have replaced 20 car parking spaces with a densely planted, biodiverse forest of some 35 native species of trees.

Other cities have explored ways to repurpose parking. We often have so much of it and so much redundancy between on street parking and spaces in parking decks. Here there is considerable potential for cities to better manage the parking they have and in so doing to free up redundant spaces that could then be devoted to trees and forests.

A recent study analyzing parking in Melbourne, Australia, provides insights into this potential. Here an analysis of the excess and under-utilized parking spaces in parking decks suggests that many on-street spaces could be considered "redundant capacity" and could be re-purposed for nature and trees.[7] The researchers found some 50,000 off-street spaces in the city with high vacancy rates. According to their calculation there could be more than 11,000 "redundant" on-street spaces, amounting to a potential (on the high end) of more than 24 hectares of city space that could be "depaved" and if planted with trees could, when those trees were mature, provide nearly 60 hectares of shaded canopy. "These changes," the researchers conclude "would also represent a large contribution to the city's ambitious '40% by 2040' target for tree canopy cover on public land, delivering up to a third of the required change."[8]

Urban parking decks as these examples show could help to free up space in cities for more trees. These structures can themselves be planted, of course, and there are today many examples of green parking structures. Years ago I encountered an elevated parking garage in Sydney, Australia, with many trees and vegetation on the top level. More recently, an elevated parking garage in the Indian city of Mumbai, has boldly pursued a similar design. Referred to now as "The Rock," it was envisioned by its designer, architect Shyam Khandekar, not so much as a place for parking cars but as, in his words, "a sort of rocky outcrop with trees and other green landscape."[9] The project renderings make the parking garage look like a green magical landscape, something to be hiked rather than driven to. With recent construction the photos are less impressive, but over time as the trees and nature grow-out it will likely slowly move closer toward those verdant images.

Forested Pedestrian Spaces

Fast forward to the global pandemic of the first years of the 2020s when many rediscovered nature and the out-of-doors in cities, and a new model of cities deemphasizing the prominence of automobiles emerges. Called the 15-minute city (and discussed briefly in an earlier chapter), Paris has been at the forefront of implementing this new idea: the vision of a city where all our daily needs could be satisfied—work, shopping, recreation—within a 15-minute walk, or a bike or scooter ride. The time it seems is ripe and many residents of cities are ready to do away with their dependence on car and car mobility. And in making this shift is the possibility of more room for trees and forests and for the possibility of enjoying them in ways other than from the window of a moving car.

Another important example, the BeltLine, in Atlanta, which has a critical tree dimension to its story. An idea originally developed and proposed by an architecture student at Georgia Tech, the BeltLine envisions when finished a 22-mile walking, cycling trail that follows the old railroad lines that encircle Atlanta. Much of the vision has already been brought to fruition and the BeltLine is already very much an attraction and an amenity loved and used by Atlantans.

Trees represent an interesting dimension to this multifaceted space. Greg Levine, co-executive director of the nonprofit Trees Atlanta, in an interview told me the genesis of what would become the BeltLine Arboretum. Returning from a leave of absence, and in the early strategies of the BeltLine, he describes being asked by then-executive director and founder Marcia Bansley what role the organization could play in this exciting new project. It occurred to Levine that Atlanta did not have an arboretum and that perhaps the BeltLine could be the location of a unique kind of arboretum, a linear version. "We really do not have an arboretum in the City," he thought. "We should create the longest arboretum in the

FIGURE 8.2 A Segment of Atlanta's BeltLine. Photo credit: Tim Beatley.

world."[10] They raised funds for a master plan for the arboretum and hired the Seattle-based company called Portico.

Today the BeltLine Arboretum is an impressive collection of trees and nature that visitors enjoy on a walk, a run, or a bicycle ride, with about ten miles of it planted. It has become a showcase of native species (Figure 8.2).

What is different about the BeltLine because of the presence of trees and nature? "You see seasonal change," Levine notes, and there is "fairly abundant bird wildlife, and at night other wildlife as well." There is a connection to nature here, not only visual but also the sounds of nature, that result from the trees and plantings.

Levine notes the high number of visitors who experience the trees and plants—nearly two million visit the eastside trail each year. And it is also a chance to educate and influence the planting decisions of many of the individuals and companies that have property on or near the BeltLine, like the Ponce Market, which Levine points to as a great example of the use of majority native plantings on its site.

"We want to use the BeltLine not only to connect people to nature, but also be able to bring part of the BeltLine home with you by having that similar plant palette and that diversity idea ... to bring it back to neighborhoods and redevelopments across the city." This has happened, he says, in a number of places. In this way the educational impact of the arboretum is magnified as many thousands of people see and experience native species of trees than might in a more distant, stand-apart conventional arboretum.

The High Line Park in New York is a dramatic example of how our view of parks and civic space changed. Many cities have been inspired by this example and have pursued similar re-purposing projects and similar linear, elevated parks. Seoul's new "skygarden" is another similar example, a former highway overpass that has been converted into a nearly one km-long elevated walking path and arboretum, containing some 24,000 trees and plants.[11]

There has been an explosion in new thinking about parks and new examples of leftover or forgotten spaces that have productively and creatively been converted to parks and community greenspaces. These have included the Underline in Miami, where a linear park has been created under that city's elevated metro transit lines, and the Bentway, in Toronto, with park space (and even a skating rink in winter) underneath the a highway (the Gardiner Expressway). These new parks and greenspaces emphasize walking and biking, and they often include extensive trees.

From Freeways to Treeways

Surface roadways and car spaces can be recaptured for nature and trees, and there are increasingly good examples of what is possible.

The Texas Trees Foundation has been spearheading another kind of example of how trees could supplant or at least soften the culture of car dependence. In Dallas, a bold design process has been underway in Southwest Medical District of Dallas. Spearheaded by the Texas Trees Foundation, a multi-years process has been underway to significantly green the Harry Hines Boulevard that runs the length of the district. Referred to as the Green Spine, it is radically reimagining this roadway as a highly traffic-calmed, heavily forested green corridor.

One part of the corridor will become the Green Heart, a kind of central park, that will allow car traffic to move through it but will also create extensive pedestrian spaces, again with many trees, in a park-like setting. The renderings suggest more trees and forests than cars. It remains to be seen what the spaces will eventually look like but the trajectory seems to suggest the need to rethink a medical district where trees and nature must be enlisted as co-healers.

In my own home city of Charlottesville, Virginia, there is a wonderful example dating back to the 1970s of the pedestrianizing of the city's main street and the planting of relatively larger, mature trees. It is one of the most pleasant walking and sitting environments anywhere, made possible by the planting of the trees. Specifically, the pedestrian mall can boast more than 60 large trees, mostly willow oaks, planted not in a line but in clusters. As a result, much of the mall is shaded in summer, and at least ten-degrees (Fahrenheit) cooler than surrounding streets and spaces. It is common to see entire families walking and strolling the length of the mall, and especially on weekends it has become a haven for buskers and artists of all sorts, and a beehive of energy and activity.

Much of this was by design and the work of the American landscape architect Lawrence Halprin, known for his designs for Ghirardelli Square, in San Francisco, and the FDR Memorial in Washington, DC. His work often involved water, trees and connections to nature, all important elements of the Charlottesville Mall. He saw these public spaces as essential to the vitality and life of cities. They were stages "set for human drama," as one profile of his work called them, providing a "choreography of seren-dipity."[12] These are public spaces that draw people in, that make them want to sit and stroll; at once natureful and urban, and the trees, the small groupings of trees, were an essential reason for this.

Urbanist and researcher of the Downtown Mall, the late William Lucy was a keen admirer of the trees and believed that their size and number were a big part of why the Charlottesville Mall has worked (when many of the pedestrian malls of the 1970s were eventually taken out in US cities and returned back to auto traffic). "The number and height of the these trees," Lucy believed, was an essential part of why the Mall works: "they create an atmosphere of natural beauty in a dense built environment that is aesthetically pleasing ... [that] con-tribute to a congenial microclimate with low pollution and pleasant bird sounds."[13]

Few of even the best European spaces can compete with our Downtown Mall. One that does is the main pedestrian plaza in the Lithuanian city of Kaunas. Visited more than a decade ago here, the Laisves Aleja contains many trees planted in two parallel rows work to create a shaded prome-nade that extends for about 1.6 kilometers. It is another form of a treeway (Figure 8.3).

Arlington, Virginia, another pedestrian-friendly city provides another kind of treeway example. Here, the city has committed to the vision of biophilic urbanism, and recently that has taken the form of what it calls the "Green Ribbon." The key organizing idea of its recently released *Pentagon City Sector Plan*, the Green Ribbon, is described as "a connected network of generous, biophilic walking paths."[14]

FIGURE 8.3   The Laisves Aleja, Kaunas, Lithuania. Photo Credit: Tim Beatley.

Forested Caps and Rooftops

Especially in the American context, there is underway a major rethinking about highways, and many cities are taking steps to retrofit and remedy the ill effects of this history of highway construction. This movement to rethink highways is also a major opportunity to plant more trees and expand the urban forest, though the potential for urban reforestation has been a secondary consideration in most cases.

Rochester, New York, has famously been a leader here, closing portions of its inner loop and attempting to restore neighborhoods that were separated as a result of this highway. In a recent *New York Times* article reporting on efforts to plan for further dismantling of the Rochester highway a resident is quoted as favoring the new plans in part as a way to add more parks and nature to the neighborhood, trees and nature destroyed when the original highway was built. Marketview Heights neighborhood resident Nancy Maciuska says, in the article, she favors these kinds of improvements when the highway eventually comes down: "So people can raise their families and enjoy Mother Nature."[15] A local advocacy organization, Hinge Neighbors, has been helping with this process along.

There are by some counts as many as thirty American cities engaged in some form of dismantling highways or placing caps or lids or deck parks on them. And there is new political and financial support from the federal government to do so. The potential to place a cap or a lid on a highway and to create parks and urban forests above can be seen in some of the earliest of these projects.

The Freeway Park in Seattle is believed to be the oldest example, built in 1976 over a section of Interstate 5. Another famous design of Lawrence Halprin's (with Angela Danadjieva), it boasts some remarkable park spaces, including not only waterfalls and dramatic public sculptures but also plenty of trees. A more recent dramatic example of highway lid (or deck park) is the Klyde Warren Park in Dallas. This 5.2 acre park includes some 322 trees and has, since it opened in 2012, become the city's "beloved town square."[16] More than a million people visit this "urban oasis" each year.

One of the largest examples of a deck park can be found in downtown Phoenix. Here, the Margaret T. Hance Park has been built onto a large section of Interstate 10. The park is a remarkable 32 acres in size. It is in the midst of a $100 million revitalization, led by landscape architects HargreavesJones, the renderings suggest an even more verdant set of public spaces including many trees.

Other cities have taken similar approaches. In Atlanta, a city of many freeways, there are at least three major proposals to cap different parts of the city with significant private funding recently announced for one of these. Atlanta has been a poster child of a city heavily dependent on highways, and a history of highway design and construction explicitly intended to separate and segregate. There is an opportunity in Atlanta and elsewhere to address through such freeway capping projects tree and park equity and this is one reason there has been growing public support for such projects.

The idea is finding support in many cities beyond the US. In Barcelona, for example, a major linear park has been approved—the so-called Green Diagonal—that will create much new park and nature space on top of and over roadways, though it has not yet been constructed.

It is possible that more space for trees could be created through the design and construction of land bridges and wildlife passages, something many cities are beginning to invest in. Bridges of all kinds, perhaps most especially those that connect existing parks and natural areas, present special opportunities. Many years ago, I had the chance to visit the "green bridge" in London, a project that created a vegetated foot bridge in the Tower Hamlets borough. Bridge parks are rising in popularity. The 11th Street Bridge Park in development in Washington, DC, that will span the Anacostia River, or the proposed Richmond Bridge Park are examples.

Increasingly we are reimagining building rooftops as places for nature, to plant not only meadows and native plants but also trees and forests. One example is Philadelphia's *Cira Green*—a private greenspace on the top of a twelve-story parking garage with spectacular views of the skyline of that city. Recently profiled in the *Living Architecture Monitor* there is explicit mention of the design of this space to accommodate trees.

Opened in 2015, a main benefit and function of the park is the collection of stormwater and it does this very well—retaining some 700,000 gallons per year.

> Trees are a quintessential part of urban parks, and Cira Green is no exception. Nor should it be—elevated landscapes offer an intriguing opportunity to increase our urban forests in light of the common challenges facing street trees. The design placed particular emphasis on appropriate soil volumes to support shade trees. This is most visible in the raised planters at the edge of the park. It is also true however for the Yellow Oaks growing in the high ground below the sloping lawn. In both these settings, the soil volume and structural design is sized for mature tree size, and to accommodate the structural root zone of the fully grown tree.[17]

It is a lively space and even though it is privately owned and managed there are many public events occurring there from concerts to dancing. The trees are a modest part of the park to be sure, with much of it in the form of open grassy space, and an emphasis on native species of meadow plants.

Many cities are now subsidizing and supporting the installation of green and ecological rooftops. And some cities, including San Francisco, Toronto, and New York, mandate some form of green rooftop for new buildings. San Francisco and New York mandate either a green rooftop or a solar rooftop, or some combination of the two. Mostly these are extensive green roofs, with shallow soils and sedum plants, but some do incorporate trees. The potential for trees and forests on rooftops is considerable. It is estimated that while New York City already boasts some 730 green roofs, though they together take up but a small fraction of the amount of rooftop space available, for trees and nature but many other functions. It's estimated that these existing green rooftops comprise only about 60 acres (of the some 40,000 acres of rooftop space to be found in the city).[18]

In Rotterdam, there has been a concerted effort to stimulate public discussion about what is possible on its largely forgotten roofs and rooftops. The city has cleverly sought to help the public see rooftops as a new frontier for city spaces. During the summer of 2022, from late May to late June, the city organized an event called "Rotterdam Rooftop Walk," during which residents could visit and see some of these underutilized rooftop spaces, via elevated pedestrian walkways, painted a bright orange, and including a pedestrian bridge connecting to the roof of one of the city's most prominent department store. Local design firm MVRDV has been a key partner, producing the year before, a Rooftop Catalog of 130 different ideas for how these rooftops, what they have called "the second layer of the city," could be used.[19]

Rotterdam has already seen the installation of a number of green roofs, part of its approach to resilience and to managing water in the city. Some of these rooftops are already seeing trees and small forests. These include a roof of a hospital that includes nearly 40 fruit trees, and the newly opened "Depot Boijmans van Beuningen" an unusual art storage facility housing more than 150,000 art objects and open to the public,[20] and including a rooftop that includes a small forest of 75 birch trees and 20 pine trees.

San Francisco is another city supporting green rooftops. Here there is the prominent example of the Salesforce Park occupying the roof of the Salesforce Transit Hub. Extending the length of four urban blocks, it is a park that sits on the roof of a transit center some 70 feet in the sky. It has quickly become one of the city's most popular parks. As the opening page of the Salesforce Center says; "Sometimes the best surprises are right at the top. And that includes a public urban park in the sky."[21] According to this site, the park contains 600 trees and 16,000 plants. There are spaces for sitting and for enjoying concerts (there are many), concerts, as well as a perimeter walking trail. The trees are organized into different botanical zones. There are California native trees there—including California buckeye and Monterey cypress—but many tree species that are not native.

Streets as Forests, Treeways in the Sky

I agree with the opening sentiment of this chapter captured in the Simon and Garfunkel song lyric: I'd certainly rather hang out in a forest than a street, at least a street with lots of traffic, noise, and hard pavement. Streets are a key building block for cities to be sure, but they need not be only or primarily for cars. Maybe the street should be reimagined as a forest!

The last decade or so has indeed been a very fertile time in urban planning for rethinking and lots of wonderful pilot projects and experimental initiatives have sought to readjust priorities away from cars to pedestrians and community uses. All these ideas to some degree or another help to create more space for trees. Many cities have adopted complete street policies which can and often does make more room for street trees. Various concepts of shared streets (or woonerven, as they are called in the Netherlands) offer the prospect of more space for trees. Barcelona's now famous idea of "superblocks," where car through-traffic is banned and where car and roadway spaces become small neighborhood plazas are also spaces where more trees can be found, providing important shade that make these community gathering spots hospitable and attractive. There is a proposal in Copenhagen to convert a major road into forested pedestrian space. With a design and wonderfully evocative renderings by the Danish firm Gehl, it would be a major transformation there.

In Singapore, an extensive network of pedestrian trails and walkways, many surrounded by trees, allows one to travel easily to parks and larger greenspaces. The city's Park Connector Network now extends to 350 kilometers in length (Figure 8.4). One of the most dramatic segments is the 10-kilometer long Southern Ridges trail. Much of this takes pedestrians off the street and through the tree canopy on an elevated metal walkway.

In hilly Pittsburgh, one is always surrounded, by trees and forested slopes and hillsides. This topography has historically been a challenge for re-sidents, leading to the development of a network of 850 sets of city steps (or stairs). The city steps are a direct response to the distinctive topography of the city and have been a way for workers living in the relatively steep neighborhoods to travel to their jobs in steel mills and factories along the city's rivers. While other cities have similar public stairs, Pittsburgh's are unique both for their number, and for the extent of the engineering and construction that has gone into them—many include elaborate switchbacks on elevated concrete pilings. (see Figures 8.5 and 8.6).

While many of these city steps are in disrepair, the city has in recent years developed a plan for identifying the most important ones and gradually repairing and sometimes replacing these structures. The steps are an unusual and unique feature of Pittsburgh, cultural and historic assets, and certainly an important part of the pedestrian infrastructure of the city, but they are also important connections to the city's hillside trees and nature. The steps provide access to steep forest patches all around the city.

There are a remarkable number of trees nearby to the city's steps and growing in the small parcels of green and open land surrounding the steps. The importance of these trees and forests that occupy the steps, slopes and largely unbuildable lands throughout this hilly city are not to be under-estimated and for the most part there are otherwise few places to access them. You see them on the hills around you wherever you might be standing in the city. In many cases, the city steps deliver you to a point where you can nearly touch the trees around and the experience of walking up or down the steps is often a highly natureful, even nature-immersive experience.

FIGURE 8.4 A Segment of Singapore's Park Connectors Network. Photo Credit: Tim Beatley.

FIGURE 8.5   Pittsburgh City Steps. Photo Credit: Tim Beatley.

FIGURE 8.6 Pittsburgh City Steps. Photo Credit: Tim Beatley.

Many of these small greenspaces located throughout the city are hard to maintain and the city lacks the staffing or budget to fully manage these forested and vegetated lands. One creative approach has been use herds of goats from a local company called Allegheny Goatscape. There are four teams of goats working sites around Pittsburgh. They arrive at a hillside location along with a guard donkey, who projects and looks out for the safety of the goats as they remain on site eating just about everything (protecting against coyotes). It is a low-tech solution to both the challenges of the local topography and lack of human staff to do this important work. And there are also side benefits, as kids (and adults) love to see and interact with the goats.

It makes me wonder whether cities like Pittsburgh should themselves invest in herds of goats that might be constantly moved around the city, eating and controlling invasives somewhere all the time. So great is the challenge that whatever tree and park management staff and structure exist, a healthy goat unit might really be an essential addition.

The city has organized some creative programming for the steps: there have been vertical block parties, for example, and efforts to reimagine the steps as vertical parks. The network of steps could help strengthen the social networks in Pittsburgh, help to tie the city together, as well as create many new opportunities for residents to visit trees and nature, albeit fragmented bits and pieces.

The assistant director of DOMI (the Department of Mobility and Infrastructure), the city's agency in charge of maintaining the rights of way, Angie Martinez, joined us to do some filming on a Sunday morning on the Vista Street Steps in the Spring Hill/Spring Garden/Allegheny neighborhood. She spoke eloquently of the value of these steps and especially their importance in providing experiences of nature:

> Something that is pretty unique about Pittsburgh's steps is that they run through hillsides, which means that they're running through areas that are developed. So you can be in a very urban landscape, like the corner of the street we're on now, with cars and cyclists and buses going by, and then you go up the stairs and be surrounded by greenspace, a little sort of step into nature right here in the middle of the city. And I think that's why people like our steps so much.

Martinez described on that day how important the steps have been to her personally, and how they have allowed her to explore the city and get to know many different neighborhoods, what she called "urban hiking": working her way through the city, finding the most dramatic views, discovering neighborhoods new to her.

Martinez describes how the city is fighting Pittsburgh is a very urban place, but these steps offer an opportunity literally to take a little break into nature right there in the middle of the city, says Martinez.

The city steps then are also a potentially important stage and venue for fostering community engagement and neighborhood cohesion. Community art is one way this can happen and there are now a number of public art projects that have taken place on the steps. Two of my favorites are the Oakley Street Steps in the South Slopes neighborhood and the Vista Street steps in the East Allegheny neighborhood.

The Oakley Steps are the site of a beautiful tile mosaic, with thousands of tiles applied to the 77 risers of these stairs, resulting in a beautiful and colorful image of a woman and flowers emanating from the smokestack of a factory. The mosaic artwork is so expansive that it is hard to see the entirety of it from the base of the stairs; it must be enjoyed by walking up (or down) the steps! This project has been described as a "neighbor-driven art project": with an overall design by artist Laura Jean McGlaughlin, funds for the project came from a crowdfunding campaign. She also ran several workshops early on to train citizens to become mosaic artists. In the end some 100 citizens were actually involved in the mosaic work, undoubtedly helping to strengthen the end of neighborhood pride about these steps.

These step public art projects not only enhance the beauty of these pedestrian routes but also provide the opportunity to insert some history about the neighborhood. The mosaic artwork at the Vista Street steps certainly do this. Here the mosaic created by local artist Linda Wallen (and supported by the Sprout Fund) is not only biophilic and beautiful (I am especially taken by the sunflowers and images of birds) but also includes many landmarks and references to the history of the neighborhood.

Spending time looking for steps, discovering new ones, climbing them, or walking down them not quite sure where they will lead, provides an unusual element of discovery, blending an experience of nature and the built environment, providing breathtaking views of both.

Conclusions

We are at an unusual inflection point in cities and in the history of urban planning. As many cities around the US and around the world reimagine their mobility systems—away from excessive car-dependence to more reliance on walking, bicycling, and micromobility options (e.g. bikes, electric scooters), it is possible to imagine more trees. Even taking a small fraction of the immense space in cities devoted to cars and reassigning it to trees and nature would transform cities. These types of actions will be necessary to bring about the full nature-immersive visions of biophilic cities and forest urbanism described earlier. How precisely can this happen? There are many promising ideas already underway. One is to rethink our parking requirements in cities. Indeed, as recounted here, the original meaning of the word parking had more to do with nature and trees than it does to finding sterile spaces to temporarily store our motor vehicles. We should return to this more natureful meaning of parking. And we should follow the lead of many cities that are being to deemphasize or downright eliminate parking requirements in cities. From Portland to Austin, cities are eliminating mandatory parking requirements, for instance. This will help readjust the balance of space in favor of trees and nature as will any efforts to reduce car dependence. In some cities there have already been creative programs for converting parking to trees and forests, and more will likely emerge.

There are also many ways to reimagine roadways and highways as spaces of trees and nature: what I have been calling a Freeways to Treeways campaign. Treeways can take many different forms, including forested pedestrian spaces (like the Charlottesville downtown mall), bridge parks, and other kinds of urban trails and pathways through a city (including the BeltLine in Atlanta and the concept of the Green Ribbon in Arlington) that further help to reduce car dependence. Capping and covering highways will present other opportunities for trees and forests, as will decisions to demolish or deconstruct existing highways. The special case of Pittsburgh's 850 sets of public steps or stairs is considered at the end of the chapter, admired as a unique historic and cultural asset that will both provide pedestrian access and also help connect to small forests throughout the city.

Notes

1 From the song "El Condor Pasa," found here: https://www.paulsimon.com/track/el-condor-pasa-if-i-could-2/, accessed June 22, 2023.
2 Donald Shoup, the High Cost of Free Parking, Routledge, 2011.
3 DC Office of Planning, "Public Space: A Defining Characteristic of Washington, DC," found here: https://planning.dc.gov/sites/default/files/dc/sites/op/publication/attachments/Public%20Space%20A%20Defining%20Characteristic%20of%20Washington%20DC.pdf, accessed May 22, 2023.
4 "Parking to Park," Science Museum of Virginia, found here: https://smv.org/learn/parking-park/, accessed June 10, 2022.
5 Ibid.
6 Wyatt Gordon, "From Parking to a Park: Can One Richmond Surface Lot Prove the Value of Depaving?" Greater Greater Washington, June 6, 2022, found here: https://ggwash.org/view/85219/from-parking-to-a-park-can-one-richmond-surface-lot-prove-the-value-of-depaving, accessed June 10, 2022.
7 Thami Croeser et al., "Finding Space for Nature in Cities: The Considerable Potential of Redundant Car Parking," *npj Urban Sustainability*, 2022, found here: https://www.nature.com/articles/s42949-022-00073-x, accessed May 17, 2023.
8 Ibid.
9 Shyam Khandekar, "Conceiving Parking Garage as Landscape," *Biophilic Cities Journal*, June 2023, found here: https://static1.squarespace.com/static/5bbd32d6e66669016a6af7e2/t/645d35a72b48774a67bfef0f/1683830185873/Parking+Garage_Khandekar.pdf, accessed June 6, 2023.
10 Interview with Greg Levine, August 2, 2019.
11 See MVRDV, undated.
12 Patricia Leigh Brown, "For a Shaper of Landscapes, A Cliffhanger," *New York Times*, July 10, 2003, found here: https://www.nytimes.com/2003/07/10/garden/for-a-shaper-of-landscapes-a-cliffhanger.html, accessed June 22, 2023.
13 William Lucy, *Charlottesville's Downtown Revitalization*, Charlottesville VA, 2002, p.25
14 Arlington County, *Pentagon City Sector Plan*, 2022, found here: https://www.arlingtonva.us/Government/Projects/Plans-Studies/Land-Use/Pentagon-City-Planning-Study, accessed June 6, 2030.
15 Nadja Popovich, Josh WIlliams, and Denice Lu, "Can Removing Highways Fix America's Cities," *New York Times*, May 21, 2021.
16 See https://www.klydewarrenpark.org/about-the-park/our-story.html, accessed May 16, 2023.
17 "Case Study: Philadelphia's Cira Green Project: Innovation in Urban Placemaking," *Living Architecture Monitor*, Fall 2022, found here: https://livingarchitecturemonitor.com/articles/philadelphia-cira-green-project-fa22, accessed May 22, 2023.

18 Anne Barnard, "How a Rooftop Meadow of Bees and Butterflies Shows NYC's Future," *New York Times*, October 27, 2029, found here: https://www.nytimes.com/2019/10/26/nyregion/green-roofs-nyc.html, accessed June 6, 2023.
19 See MVRDV, "So much more is possible on our rooftops than we think," May 31, 2021, found here: https://www.mvrdv.com/stack-magazine/3878/rotterdam-rooftop-catalogue-interview, accessed May 17, 2023.
20 See www.boijmans.nl/en/depot, accessed May 17, 2023.
21 See https://salesforcetransitcenter.com/salesforce-park/, accessed May 16, 2023.

9

NEW (AND OLD) IDEAS FOR CITY TREE CONSERVATION

Cities are in need of a more expansive and robust set of planning ideas and conservation tools to protect trees and forests. This chapter aims to review some of the most promising of these tools and new directions. It is an exciting time, given the many new initiatives, technologies, and organizations working in this space, and I am only skimming the surface in what follows.

While there is a long history in the US and elsewhere of conserving lands and landscapes through the use of conservation easements and land trusts, there are relatively few such examples when it comes to cities. What might cities do to protect, even secure ownership of trees and forests in existing neighborhoods?

This is a different challenge than that faced by some of the development-oriented tree ordinances that establish minimum canopy requirements. This scenario is about preserving the larger trees that exist mostly on the private property of homeowners, in front yards and backyards. These trees provide many public benefits and considerable collective ecological value, yet they are a form of private property—the right to cut them down, especially in the absence of a strong law to the contrary, is understood to reside with the property owner.

One option might be for a city to itself purchase a heritage tree, most likely by buying the underlying land that supports that tree. There are some notable examples of this approach, though this seems to happen fairly rarely. In Toronto, the city agreed to purchase a home and parcel from a private property owner as a way to save what some believe may be the oldest tree in the city, a 250-year-old red oak. The city agreed to pay

DOI: 10.4324/9781003377344-9

the owner $780,000 Canadian (or about $600,000 USD), a relatively large sum even for a city of Toronto's relative size and wealth. The idea is that the home will eventually be removed, and a small public park created out of the parcel with the tree at the center.

The city negotiated a sale with the owner in 2019, contingent upon the public coming up with a significant part of the acquisition price ($400,000, or about half). It is an impressive tree, to be sure, and one that has garnered much support for its protection. A unique aspect of the story is the community fund-raising involved, with more than 1500 individuals contributing money to pay for the purchase (and under the agreement securing $400,000 from community donations). And there were strong voices behind the public purchase, including neighbor Edith George who called the tree her "cathedral."[1] Or the former elected chief of the Mississaugas of the New Credit First Nation, Carolyn King, who sees the saving of this tree as a step in preserving the stories and heritage of her people who lived for centuries where Toronto lies.

That so many Torontoans were willing to contribute to fund the acquisition is impressive. One young activist—6-year-old Sophia Maiolo—raised $735 from her friends and family. She lived nearby and her parents frequently took her to see the oak. "She fell in love with this tree," her mother tells CBC news.[2]

A snag was hit when the owner demanded a higher sale price, arguing that this was fair given that the price of homes had increased in the city. A court ruled in favor of the city and upheld the original negotiated price.

In this case the city chose to purchase one especially notable tree, with the intent to tear down the house and to establish a small park.

Would it be possible for every city, or many cities. to engage in more extensive acquisition of trees in a neighborhood, or in multiple neighborhoods, as a tree conservation strategy? There are few examples of this though nothing (other than financial resources) would prevent it. As with the purchasing of a conservation easement or development right another option, in theory, would be for the city, or other entity, to purchase just the tree-rights or timber rights, severing these from the underlying ownership of the land.

Changing the Financial Incentive Structure in Cities?

The idea of securing tree rights is appealing in that the cost would be much reduced (compared with purchasing an entire home or parcel to protect the tree). There is considerable precedent, moreover, for it. There are now many other creative ideas for protecting city trees, some that rely more on financial and other forms of incentives.

Norfolk, Virginia, has begun to integrate consideration of trees into its larger coastal resilience framework and in this way incentivizing tree protection. Under this innovative approach the city has developed a unique points-based code that requires new development to meet a minimum "resilience quotient' (RQ). As of 2021, trees are now taken into account in two ways. For all single family homes, the RQ mandates certain basic things, including, under the element of stormwater management, the installation of rain barrels. Now, instead, a homeowner can meet this requirement by preserving existing trees or planting new trees (specifically, depending on the amount of lot frontage). For multifamily projects, preserving or planting trees is also an option, with more points given to projects that preserve existing trees.

Developers can be given substantial incentives to better protect the trees on development sites. These would be in addition to enforcement of strong tree protection codes. There is a long tradition, for example, of using density bonuses to achieve a variety of environmental and urban planning objectives. Tree protection could be part of the answer—offering generous additional density for steps, perhaps beyond the minimums of the tree protection code, that would plant more forested land, or protect more trees and forests.

There are other interesting precedents that relate to the changing perceptions of what residents should be doing, or able to do with the spaces immediately around their homes. I have recently written about the provision of a financial subsidy offered to residents of Marco Island, Florida, who are willing to host a nesting site in their yard for burrowing owls.[3] In this case, the payment is relatively small, and homeowners seem more strongly motivated by other considerations. Nevertheless, even a small subsidy for homeowners willing, for example, to keep half their lawn in trees and native vegetation would be helpful.

There are also precedents in many western US cities in the provision of financial subsidies for the conversion of thirsty turfgrass yards to more water-conserving, xeriscaping. Many cities in arid climates that are grappling with limited potable water have adopted some version of this approach (e.g., cities of Los Angeles and Las Vegas). These subsidies are usually only one-time payments. Many cities have stormwater districts, as another example, that adjust the fee depending on site conditions. The more vegetation present and the more permeable and less paved the site, the lower is the stormwater fee. If a homeowner were to install a green roof or take up some hard surfaces, the fee would go down.

Traditional local property tax systems also need major reform and could help to further stimulate tree planting and tree protection. This could happen in several possible ways. Perhaps there could be some sort of urban version of the use-value taxation provisions that many rural communities use to help protect farmland. Under these programs land-owners are eligible to have their land assessed and taxed based on its "use-value" that is its value as farm and forested land, and not the high fair market value often connected to land speculation and development. Could a suburban property owner be assessed as a forest, or partially as a forest, serving to reduce his/her tax bill.

More complexly perhaps, could a resident's overall tax bill be deter-mined, at least in part, on the ecological services and ecological contri-butions made to the larger community? Counting the number and size of the trees on a site could factor into the tax bill the property owner pays—the more trees on the parcel, the lower the tax bill would be. And this could change year-to-year as annual tax assessments do: if trees are cut down or lost, the accounting takes this into account and the ecological benefits go down and in turn the tax bill goes up.

Adjusting the property tax system to reflect the presence of trees and for-ests is a promising direction perhaps for each older tree protected, a home-owner receives a sizable deduction from their tax bill—perhaps $200 each year for smaller trees, perhaps $500 per year for larger trees, reflecting the larger community and ecological benefits provided by larger, older trees. Homeowners would (again) lose those benefits if they cut down or otherwise caused the death or loss of that tree. Or if the tree dies of disease or older age, it would be subtracted from the homeowners list of community "assets" attached to their tax assessment (perhaps the subsidy, or some portion of the subsidy, could continue if the homeowner allowed the tree to continue to stand or even to degrade on the ground, in recognition of the wildlife and biodiversity value the tree still holds).

And like use-value assessment programs (used to protect farmland) there could easily be a recapture clause included—that if the trees are cut down the homeowner or developer would be required to pay back some or all of the subsidies given (a typical payback period for use-value programs is five-years, for example).

There would certainly be resistance from the local budget office and worries about the decline in tax revenue. But there is certainly consider-able precedent for paying urban residents to encourage them to take ac-tions favorably to the larger urban environment. And there is certainly a need to go well beyond the modest and meager financial incentives for tree planting and tree protection.

The Power of Naming Trees and Forests

There is power in protecting trees by naming them. We saw this in Chapter 1 in describing efforts to save the western red cedar in a Seattle neighborhood known as "May." A similar story from Santa Fe, New Mexico, offers lessons and insight. Here, some of the most spectacular trees are the large and beautiful cottonwoods. They are a dominant species that grow to be quite large especially where there is water, for instance along the banks of the Rio Grande River in Albuquerque or along the Santa Fe River. Massive cottonwoods can be found throughout the city of Santa Fe and they are incredibly important elements of what makes this a very distinct and special city.

In 2015, a conflict arose over the very large cottonwood tree growing in Sena Plaza, one of the oldest spaces in the city.[4] There a large limb of this tree fell on a table and briefly trapped a diner there. The tree was too dangerous; the owners of the plaza argued and wanted the tree cut down.

This tree had admirers, a function of its prominent location, and many called it "Willy." The city's fairly strong tree protection code left the determination of the tree's fate in the hands of the city's planning department and in 2015, they concluded the tree was healthy and should remain. But with trees, legal protection is never certain and in 2019, the issue arose again, this time the planning department, while agreeing the tree was healthy, believed the risk to diners and the public justified the cutting down of the tree.

This story makes me sad, and in 2022, I had the chance to visit the plaza where the tree used to live. There are brick pavers and restaurant tables there now. There are other trees in the plaza and it is certainly a green and natureful place. While I would have preferred the tree be saved, I can appreciate the concern for safety, and perhaps the verdant nature of the plaza today (and younger trees growing which will replace over time the giant Cottonwood) serves to mitigate some of these concerns.

I am also impressed with the process and relatively careful (and public) deliberation that went into this decision. I wonder if there were not other alternatives to address the safety issue (perhaps using the kinds of netting used in baseball stadiums, or stronger parasols to shade and protect against falling limbs in the future?), but I am glad the fate of the tree was decided in such a careful way.

I also appreciate the embrace of the community and the genuine affection for the tree. Giving it a name helps to further foster care and love for this tree and helps to guard against thoughtless destruction. It was reported that when the time came to cut down the tree in 2019, there was both a Catholic priest and a rabbi on hand to bless the tree.[5] The tree deserved the blessing and provided years of good service to humans and nonhumans of Santa Fe alike.

There are many means and methods of expressing reverence for trees. And there is a need for commemoration and for ceremonies of celebration and thanks for what trees have given us.

Native American writer Robin Wall Kimmerer, author of the important and impactful book *Braiding Sweetgrass*,[6] describes the indigenous traditions of thinking of trees as people (referring sometimes to the "standing people") and kin. Partly this is about the language we use; she suggests the importance of a "grammar of animacy." Especially concerning is the western practice of attaching the pronoun of "it" to trees that are living kin. Such practices tend to objectify, she says, treating a tree more as an object than a living subject.

Collecting personal stories about the important trees around us and in our lives can be an effective way to personalize them. I am taken by the work of the Boston-based organization Speak for the Trees, and the personal stories they have captured through YouTube films. One is the story of Cambridge resident "Robin" who talks affectionately about the large pine tree she planted as a sampling more than thirty years ago. She refers to this tree as "him."[7] "He's a wonderful, wonderful tree," she says, clearly conveying something very close to love for this tree. "You can depend on him," she says, and explains how reassuring it is to see the large tree as she walks toward her house each day. "He's there, it's all good, it's all good."

Robin Wall Kimmerer's writings eloquently describe the Native American traditions and culture of viewing the natural world through a perspective of gratitude and reciprocity. Nature, especially trees, give so much to us, and their gifts should be acknowledged and thanked. Practices and ceremonies of thankfulness can be undertaken by individuals and communities alike and help to stimulate reciprocal actions. The gifts that trees give us—shade, water, food, birds, and birdsong among others—can be formally acknowledged and efforts taken to repay and reciprocate. What forms might reciprocation take? Watering and caring for the trees in one's neighborhood, planting new trees, appearing before the city council to ensure that older trees are protected?

When trees must be taken down it is cause for community attention, reverence and sadness, and celebration of the fullness of the life of that tree. Recently a grand willow oak had to be removed in the City of Raleigh's Nash Park. Residents came and watched and the crew cut small 2-inch sections of the trees to hand out to residents as keepsakes. In the case of Raleigh, the city's very beginnings are tied to the oaks that existed there, and that led to its city moniker of "City of Oaks."

Similarly, in many cities community vigils and celebratory events are organized when old trees are slated for cutting. This often entails an element of protest and an effort to prevent the cutting. This happened dramatically in Sheffield, England, where residents there were shocked to learn that large trees near to their homes, often along the street, were slated for cutting, a primitive and shortsighted decision by a global corporation hired to manage the street trees.

There are many creative ways to highlight the presence of remarkable trees in a city. In many cities there is some form of "tree of the year" competition and celebration. It usually depends on the nomination of a tree by someone in the community, after a call, and then some form of public voting on the candidate trees. Raleigh, North Carolina, is one city with such an annual call and designation.

In Europe, there is a European-wide tree of the year nomination and competition. This has been going on since 2011, first organized by the Czech Republic. The year 2023 saw an oak tree in Poland win with more than 45,000 votes.[8]

Another approach is to make room for aging trees, even when they fall, and are in the advanced stages of decomposition. They can still represent important community features and sites to visit and to learn about the deeper history of a place.

As mentioned in an earlier chapter, some years ago, I visited the so-called Queen Elizabeth Oak, or what remains of it, found in London's Greenwich Park.[9] Here this ancient oak remains in a state of decay, a home to invertebrates and other creatures but still very much a time-marker. The tree dates back to the 12th century, and King Henry VIII and Anne Boleyn were said to have spent time under its canopy. It died some 200 years ago, but until 1991 it was standing, then finally came down in a storm, to continue its presence and decay, and its history and life lessons.[10] The Swedish speak of their ancient oak trees in this way: they live for 500 years, then they die for 500 years. The Queen Elizabeth Oak is giving practical meaning to that expression.

I contrast that reverence and attention with the treatment of another "queen tree," this one in the borough of Queens in New York. We found it while during a documentary film shoot in Alley Pond Park. It is a few feet away from the Cross Island Parkway, hard to find and encircled by a metal fence. The tree is little known by New Yorkers but estimated to be between 300–350 years old. Gazing at it is humbling, and in the context of a highly developed city, a remarkable sight. Measured by the city in 2000, it is 134 feet tall, with a circumference of 18 feet.[11]

During the recent controversies around building a wall along the southern border with Mexico, concerns have emerged about environmental impacts. A rallying point especially for the Native American community, has been a 900-year-old Montezuma bald cypress tree. It is a stand-in for all the potential ecological destruction that a wall will bring, and a natural rally point given its age and how many past generations have known the tree. It is as if this tree is standing watch, a somber witness of posterity who might be called upon at some later date to report on what happened there.

So-called "wall warriors" of the Carrizo/Comecrudo tribe, came to this site near Mission, Texas, a place that is also a sacred burial site for their people. There they "prayed and sang songs to the grandfather tree."[12]

There are other traditions that result in making deeper time more visible. In Nordic countries, the tradition of carving runestones, that stand or jut out from the earth and are usually in honor of a parent or family member. Many are 1000 years or older dating back to the Vikings. By one estimate there are 2500 of these carved stone memorials (in Sweden?).[13] Ancient trees can serve this time-deepening function in cities.

Tree pilgrimages would be helpful and an awareness of older trees and forests around us in cities. On a recent visit to Atlanta, Georgia, I sought to find the oldest remaining tree in the city. I was helped in this quest by the work of a local nonprofit, Trees Atlanta, which maintains a list of "champion trees" in the city.[14] With the help of this list I went in search of the largest diameter, a Cherry bark oak (*Quercus pagoda*), located on Washington Street adjacent to the offices of Our Lady of Perpetual Hope.

The tree was of a remarkable size, and its crown provided shade even to parcels on the opposite side of the road (Figure 9.1). It was also teaming with birdlife. The circumference of this tree is recorded to be 276 inches, making it, according to one tree calculator, more than 350 years old. Atlanta is a relatively young city, established only in 1837, which makes the tree some two hundred years older than the city. This is a tree that has witnessed the city's entire history and development, worthy of attention and reverence, and periodic pilgrimage to see and visit it by residents. I am not sure when last anyone visited this tree, however. I was pleased to find it, to experience its majesty, and to reflect on the history and deeper time it embodies.

Such champion lists are helpful, to be sure, but what is needed is the cultivation of an ethos and mindset of paying attention, and homage, to these ancient tree elders around us.

Toward an "Internet of Trees"

We live in the digital age where IT technologies are continuing to shape our lives and also increasingly change the ways we are managing our cities.

FIGURE 9.1   The Cherrybark Oak in Atlanta: Is It the City's Oldest Tree? Photo Credit: Tim Beatley.

Can these digital technologies help us to better manage and protect urban trees and forests, and might they help to connect us emotionally to them? Kids and young people continue to bear the disproportionate brunt of the shift to a digital world. Richard Louv, in his important book *The Last Child in the Woods*, worries about the shift toward indoor spaces, the growing sense of danger about the outside world, the replacement of direct experience of nature with immense amounts of digital screen time.

But can this same digital technology help us when it comes to urban nature and trees specifically. Apps like iNaturalist definitely can help. This is an app that I use almost daily and is highly useful in identifying the things we encounter around us in the natural world, including trees. There are also apps that help us to locate the nearest parks and forests to visit and walk in, and some apps that help to remind us about how long we have been online and nudge us to take breaks and to walk outside. We can use our cell phones and computers to make plans to travel to a national park or forest and many other ways our lives have been enhanced as a result of this technology, certainly.

The mass diffusion of smartphones represents, as well, a technological trend that could not only distract from trees but also help connect us to trees. Apps like iNaturalist have been hugely important in helping average citizens to understand what they are experiencing and seeing around them in cities. The app is an opportunity to participate in citizen science by snapping photos and uploading these observations to a geo-coded database, with verification from other citizen scientists who can confirm a sighting.

The smartphone has become a tree-friendly tool in another way. Many cities are now placing QR codes on trees, allowing citizens to scan the code and learn more about the tree—its species, age, etc. Rotterdam has, as mentioned earlier, been using tree bar codes as a way to inform residents about plans to remove trees, for instance because of a road or construction project, or because a tree is dead or diseased. It becomes a way for a city to be more upfront and transparent about its plans for its trees rather than surprising residents with a tree removal.[15] Providing this form of advance notice might also give residents the chance to object or voice their opinion in some way that could meaningfully change the outcome.

In Washington, DC, QR codes on trees have been used to solicit care for trees, through a pilot program called Canopy Keeper. Here QR codes were attached to newly planted street trees. "Passersby who scan the image with a smartphone will be automatically directed toward a form on the Department of Transportation website to register their interest in caring for the tree."[16] One of the key commitments for citizens signing up to adopt a tree is watering the tree (and they receive a free "slow-drip watering tub" to make this job easier).

In the last decade or so, especially, there has also been a rise in the ideas connected to "smart cities." Here the emergence of the new and inexpensive sensors has led to many useful ideas for their use, many having to do with managing traffic and transportation in cities. Cities can now manage a variety of functions and management tasks more efficiently, for instance sensors in some cities that can indicate when a trash can is full and in need of emptying, or sensors that provide real time information about flood conditions in a city (which streets are flooded and which are not), aiding more effective decisions about evacuation. The revolution in urban sensors holds much potential to enhance the quality of life in cities certainly, as it makes urban services more efficient.

Sensors are also being creatively and extensively used on behalf of nature and nature consecration. Recent work by a UVA professor, for example, has led to the development and piloting of a sensor to help determine when endangered sea turtles are about to hatch.[17] It is cleverly disguised in a fake turtle egg. An ornithologist colleague has been developing a sensor for windows that will help detect when and where bird strikes occur, and with the potential to lead to window treatments that reduce bird mortality.

There are already many examples of the use of sensors in trees and forests, though more commonly to collect data in places outside of cites.

Some tree sensors that detect motion can help to alert authorities when a tree is being cut down and thus could be an important tool for tree and forest protection.

Despite the benefits and utility of these devices, there is nonetheless push-back to report. Some are concerned about the privacy implications of smart cities where surveillance and monitoring are ubiquitous. Others suggest that the goal of an efficient city, which seems the primary task in the smart city, will not bring us the joy, delight, wonder and meaning we want and need from cities. Maybe we need cities that are just "smart enough," as one recent book suggests?[18]

The ways in which a connected city allows for communication between and among its many pieces and parts offers additional benefits. If a weather-monitoring app can predict rain and communicate that to a tree-watering device, overwatering might be avoided and precious water can be conserved. This is the future possible from the so-called "Internet of Things," (or IoT) where even common household appliances have Bluetooth capability and are connected to the internet. It portends, for some, scary futuristic scenarios—where the household refrigerator records when the milk and eggs have run out, directly ordering fresh supplies from the local grocer. This is to others an example of the convenience provided by the emerging IoT.

There are now many positive examples of what has been called the "Internet of Trees." There are "accelerometers" placed on trees on the University of Colorado-Boulder campus, for example, which has been referred to in the popular press as "tree fit bits."[19] These can be used to determine the precise time a tree flowers buds or a tree drops its leaves.

Singapore has taken the idea of an online tree map, described in an earlier chapter, to the next level. It has created "digital twins" of some two million trees found in this city-state (a large percentage of its seven million trees in total). This allows the tracking of growth over time and detection of when a tree might need to be trimmed. As a recent article in *Tech Radar* notes: "The data produced by the digital twin can be modeled to spot when urgent works are needed, cutting down on wasted resources and manpower whilst keeping safety high."[20] Many of these trees have also been equipped with "tilt sensors" that may give advance warning of hazardous trees. The city is also using satellite imagery to monitor the health of trees, allowing them to "spot trees with a lower chlorophyll rate, showing up as yellow or brown, alerting that they need immediate attention."[21]

More broadly, Nadina Galle and her colleagues have imagined an "Internet of Nature," where a variety of current and emerging digital and other technologies could be used to better protect and manage nature in cities. The rise in a number of new technologies go hand in hand with the sensors. Prominent among this new technology are a variety of robots that might aid in more efficiently managing forests, though it is less clear what these will mean for cities.

Drones have already emerged as a highly useful tool for forest managers and this will likely grow in the future. Drone-acquired imagery has the potential to enhance forest mapping and monitoring.[22] More forest managers will likely be expected to get their drone pilot's license. Not only do drones have a variety of monitoring uses but they also have been used to plant new trees. A company called AirSeed Technologies has developed a drone with a pneumatic gun that can shoot up to 40,000 seed pods into the ground per day.[23]

It is hard to know what the future holds, but there are designers working on prototypes of new technology that could be highly useful in urban settings. Israeli industrial designer Segev Kaspi, for example, has designed a suite of three "robot rangers" aimed at forest management.[24] They are odd looking robots—one for monitoring the forest (a drone), one for cutting and trimming (ground based), and finally one for planting new seedlings. It is hard to imagine this trio of futuristic robots working together in the forest anytime soon, but perhaps this is the future. I welcome a robot like "Dixon," as Kaspi named her, designed in theory to plant and seed new forests efficiently and continuously in an urban setting. Though as with any new technology there is a tradeoff: robotic tree planting takes away the possibility of humans engaged in this conscious step of committing to the future.

Relatively straightforward uses of simple apps and sensors could accomplish much. Parking space apps that might allow a city to reduce the amount of parking it needs, opening up more spaces that could be planted. Water and moisture sensors can be used to ensure survival of newly planted trees in the face of increasingly hot and dry summers.

The trick will be to utilize and fully take advantage of the technologies that will benefit urban trees and forests while being on guard not to over-embrace these technologies, recognizing their potential downsides (privacy concerns for example). There is little question that the further acceleration in use of AI, robots, drones, etc., is a trend that holds considerable implications for canopy cites.

Thinking Beyond Street Trees

In many cities trees are understood to be located primarily along streets. And as we have discussed in an earlier chapter there are many new and creative ways we need to reimagine streets, the ideas of green, wild and forested streets. But there is a need to go beyond the vision of street trees and to look for ways to expand and extend the location of trees and forests in cities.

This can happen by shifting to planting strategies where we plant trees not in rows but in larger clusters, more in line with the new ideas of the transgressive forest. A tree planted on a street corner could be redesigned with a greater ambition. It could be, as New York ecologist Marielle Anzelone says, that instead of thinking in terms of a single tree, that we design, plant and maintain these street trees as "biomes." Some small examples of this can be seen in the city of Paris in the practice of giving over the spaces below a street tree to planted gardens. I have been calling these "tree gardens" because it seems such an apt descriptor—not only trees but also efforts to grow understory plants and bushes in the spaces around the trunk of the tree.

The more we expand the palette of locations and places where we can plant trees and establish forests the more opportunities there will be to expand their coverage beyond the street's edge. Planting trees and forests on building rooftops, on balconies and terraces, on converted parking areas, and so on, will all present the chance to expand the urban forest in ways that will complement trees along streets.

Moving beyond street trees will also require us to think more in terms of urban *groves*, and doing what we can to identify existing groves or patches and to put them on maps and give them some degree of protection and status. Some local tree codes offer specific protection for groupings of trees or designated "groves." We must begin to recognize the need for greater ecological connectivity in cities and work to develop strategies and tools for connecting the existing patches and smaller forests that exist in cities. Many cities, including San Francisco, Edmonton (Canada), and Singapore, have all emphasized the importance of ecological connectivity in their community planning efforts. Ecological corridors and connections must ideally extend beyond city boundaries to encompass larger regional and even continental landscapes. City trees and forests should increasingly be seen in the context of these larger ecological systems and landscapes.

Tree Rituals and Reverence: Commemorating Life Events

It is sometimes observed that residents of modern western societies do relatively little to acknowledge and commemorate significant life events. Birthday parties sometimes, maybe an anniversary or an occasional retirement. We would benefit from more rituals in our lives and trees are an unusual opportunity to mark the major events of life.

Trees—the planting of new trees, the protection and reverence of existing trees and forests—represent unusual opportunities to create new connections to place and to add a layer of meaning to life. Planting a tree is an especially powerful symbol of our commitment to the future, of course. And the step of planting a tree can provide a moving tribute to an event or person.

Elon University in North Carolina started an admirable tradition of giving acorns to incoming students, and then when they graduate an oak sapling. The traditions started in the early 1990s with a graduation speaker from the timber industry who gave every student a redwood seedling. The tradition makes sense especially here, as the word Elon is Hebrew for "oak tree."[25]

As news stories over the years attest, the acorn and seedling tradition there is quite meaningful, and students take seriously the question of what to do with their young trees. Some end up in pots, others planted in the ground, but they end up in various places around the country.[26] But it does seem to stimulate reflection on one's growth and development over time (from acorn to seedling and beyond), and the role of the university in helping to bring this about. And it is a tangible and potentially long-living connection back to the university. In 2017, a planting ceremony was begun for foreign students, who have a harder time taking their seedlings home. To this end they have created an International Grove on their South Campus as a place for planting these trees.[27]

I like very much the idea that even small college graduation ceremonies entail not only the chance to reflect on and appreciate trees, but also a tangible commitment to our collective future. Elon University has a graduating class of around 1500, so each year a not insignificant forest is being launched. Larger universities could have an even greater impact, and perhaps like what is done for the international students, each year might represent the commitment to receive a sapling (of maybe a sapling of a species of your choice from a diverse mix of native species) and to plant it in a place that becomes a future forest.

I started to do something similar (before I had heard of the Elon traditions) in several of my University of Virginia classes. The acorns I gave out were collected from our grand white oak tree, the largest tree around us in the UVA School of Architecture (Figure 9.2). It is a steady presence, a symbol of strength and endurance, and an element of beauty that elevates and uplifts our daily lives. I saw the acorns as physical embodiments of human potential and of a flourishing future, especially meaningful for overstressed students to ponder.

All of these acorns had already germinated, and so I thought there was a good chance that they could be successfully planted in a pot, and could be cared for and enjoyed for some of the time students were at the university.

FIGURE 9.2   A Germinating White Oak Acorn—One Given to My Students. Photo Credit: Tim Beatley.

Modern gift-giving feels to many a hollow, meaningless gesture. What is purchased is frequently an object, often made of plastic, that will likely eventually find its way to the landfill. What if we could turn those many gift-giving moments into the opportunity to rather than degrade the environment, to actively restore it.

Sami Grover, writing in *TreeHugger*, shared her practice of giving her two daughters the gift of trees on birthdays and other occasions; not actual trees, but a contribution to the nonprofit *One Tree Planted*, which plants trees in countries around the world. The gift would appear in the form of "hand-drawn trees" in the daughters' "forever journals."[28]

I like this "holiday tradition" very much, and organizations like One Tree Planted make it easy to gift trees and forests. It is possible even to specify where you would like the trees planted and OTP is operating on a global scale. One Tree Planted has experienced an impressive and sharp upward trajectory, planting more than 23 million trees in forty-two countries in 2021 alone (and some 40 million trees since it was founded by Matt Hill in 2014).[29] They work with partner organizations on the ground, involving thousands of individuals in the actual planting of the trees.

There are drawbacks of course. One is the inability to see the trees that have been planted or even to be sure that they have. And how do we know these planted trees will actually survive beyond their young sapling status and who will take care of them? One Tree Planted is increasingly focused on monitoring and on training local partners to use new monitoring technology such as drones, that might provide more assurance that the trees will survive (or be replaced). But there remains the impediment of spatial disconnect—those donating to plant trees in Kenya or Peru will not likely be in a position to see these trees firsthand and to object when they die or are insufficiently cared for.

And while there may be a family connection or other reason to plant trees in Colombia or Mexico, it might be more effective as a place-strengthening step to be able to contribute to tree planting locally or regionally. There are certainly going to be opportunities in some cities to do this but not enough organizations provide this option.

Reimagining birthday gifts in the form of tree planting is an excellent idea in part because it at least has the potential to fund the growth of forests over time, something that in theory could be tracked. If the parents of a newborn decide to plant 100 trees each year as a birthday gift, by the time the child graduates from college there will (in theory) be a forest of some 2000 trees, many having (hopefully) grown beyond seedlings to form the beginning of a mature forest ecosystem. It is not only a small investment in the earth but also in the mental health and happiness of the child, who will (might) revel in delight at the knowledge that there is a forest out there that owes its existence to their birth.

Christmas is of course a natural holiday to celebrate the wonder and magnificence of trees. It is not clear, however, that the practice of buying a cut tree sourced from a tree plantation helps much to restore our global forests, and there are better, more creative options in some places. One of my favorite ideas is the option of leasing a Christmas tree. San Francisco's

Environment Department in collaboration with the nonprofit Friends of the Urban Forest, offered this service for a number of years. Their Green Holidays Trees offered residents of the city (for the price of $95) the chance to lease a living tree that later would be planted somewhere in the city. "After they green your holidays," says the city's website, the trees "will provide environmental benefits in San Francisco by capturing carbon, cleaning the air, reducing stormwater runoff, and providing wildlife habitat."[30] The live trees are delivered to one's home and then, after the holidays, will be picked up for planting somewhere in the city. A clever idea, it takes the money that would be spent on a cut tree and essentially uses it to help fund the city's forest.

The Christmas tree leasing option is one offered as well by a number of private companies. San Jose Christmas Tree Rentals has been offering this option commercially for more than twenty years.[31] They will deliver and later pick up a living tree of 6'-7' in height for $200. You get to keep the tree for twenty days before it is picked up and returned to the nursery. It is even possible to reserve the same tree each year.

Birth especially is an opportunity for joyful recommitment to place and to strengthen bonds with the natural world around us. There are a variety of traditions related to the placenta. In some cultures, the placenta is believed to be a guardian angel, in some the child's twin, and in others a protective jacket the child retrieves later after death.

In many indigenous cultures birth is intimately connected to trees, for instance through the practice of burying the placenta under a tree. The placenta might be buried at the same time as a tree; the tree then becomes a protective force for the child in life. The Djab Wurrung people of Australia have had a tradition of planting a tree when a child is born—and the tree then becomes the guardian and source of wisdom in life. "The child's placenta was mixed with the seed and from then on, as it grew, it became the child's own 'directions tree,' a place where they could come for spiritual guidance."[32]

In Navajo culture, burying the placenta was a way to essentially secure a child's connection to place. "In the Navajo tradition, burial of the placenta within the boundaries of the child's tribal land will bind or root the child's spirit to his ancestors and to the land," says Sarah Hollister in a summary of placenta burial traditions. "The Navajo believe that this will ensure that the child will always return home."[33]

There has been a resurgence of interest in these placental traditions, and in shifting away from the more typical Western view of the placenta as a kind of medical waste to be disposed of. Writing in a recent essay in *The High Country News*, recent Wyoming mother Nina McConigley discusses her decision to bury her placenta under a Juniper tree in the hopes it will bestow on her daughter a deep sense of home. It seems for her a partial antidote to her own upbringing and life of transience.[34] Many Americans have similar feelings and burying a placenta under a tree, or planting the tree and placenta together may provide a deeply satisfying ritual and create a lifelong sense of connection to place, as well as an especially meaningful part of nature to care for and periodically visit.

Death is another important milestone and a chance to commemorate life through the planting of a tree or the conserving of an existing tree or forest. There are now a number of new options for tree planting and tree conservation connected to the death of a loved one, indeed part of a larger cultural reconsideration of modern burial methods.

The rise of green burial options has presented new opportunities to support forest and nature conservation. According to Lee Webster of the Green Burial Council, there are now some 220 green cemeteries in the US alone.[35] It has been partly motivated by a sense of discontent with the usual high costs and stressful aspects of the usual casket/graveyard approach. In response there has been a gradual rise in the popularity of cremation. But there are other ideas taking hold, including the notion of being buried without a casket, directly in the ground, allowing one's body to decompose naturally, melding with and sustaining the earth and the plants and trees on a site. Several years ago I took a tour of one such place—Duck Run Cemetery near Harrisonburg. While not a fully forested site, it was much more natural and natureful than a typical cemetery.

There is an even bolder version of a green cemetery, an idea known now as a Conservation Cemetery. Here, the cemetery and the burial of the dead are but part of a larger agenda of landscape conservation and restoration. Examples include the Ramsey Creek Cemetery in South Carolina, and Carolina Memorial Cemetery near Asheville, North Carolina. These follow established conservation principles and go further to protect and restore the landscape, often through a relationship with a local land trust. The Carolina Memorial Cemetery, as an example, is a site of natural forests and meadows, where bodies must be buried only in biodegradable containers, such as a cardboard casket. The cemetery's site declares: "Preserve, protect and renew the land with your burial."[36] The cemetery has undertaken a variety of impressive ecological restoration projects funded at least in part from the burial proceeds. The site is open to the public and includes walking trails.

There is another connection to trees and forests in the typical products that go into conventional burials. These include wooden caskets, which according to the Carolina Memorial Cemetery site, amounts to the loss of 77,000 hardwood trees (not to mention the large amounts of concrete and steel needed for conventional burial vaults).[37] Shifting away from conventional burials may help to conserve forests in this way also.

Partly of course this is about reimagining what a cemetery is. Cemeteries are already often quite natural, home to wildlife, full of greenery. They are often more like parks and many essentially serve as essential greenspaces in dense, developed cities. There is a trend in the direction of injecting even more wildness to these spaces. There is the example from the Calvary Cemetery in St. Louis, a cemetery where a remnant native prairie has been preserved and restored, even through use of periodic controlled burns.[38] Or the famous Père Lachaise cemetery in Paris, where notables Jim Morrison and Oscar Wilde, among others, are buried. Partly as a result of banning the use of pesticides, the cemetery has become home to foxes, birds, and other critters. Many now visit the cemetery for the nature there, and as a recent *New York Times* profile notes, the nature and the greenery may make the spaces less scary, and may help to "distract attention from death," as one visitor noted.[39]

This is a good trend and suggests that urban forests may be the places where in the future we go to visit a grave and to remember and celebrate the life of a friend or loved one. Or it might also be a place to go bird-watching, as now happens at Père Lachaise. More human and nonhuman life is a positive and having more reasons to visit such spaces is good.

Lee Webster suggests there is value in moving away from the notion of cemeteries as places that are sterile and off limits. This is the old model, seeing cemeteries as places "with perfectly manicured lawns with pesticides and herbicides and everything else, where "you showed up on memorial day and that was it."[40] Indeed a number of the most beautiful cemeteries date back to the mid-1800s and the rise of what have been called the "rural cemeteries" or "garden cemeteries" movement. For landscape historians Mount Auburn Cemetery, west of Boston, was the first, and set the direction. Here the Massachusetts Horticultural Society set aside seventy-two acres of "natural mature forests and ponds, planted additional trees and constructed winding pathways that followed the natural curves of the landscape."[41]

Many cemeteries are often the sites of impressive trees and woods, which makes complete sense (Figure 9.3). And tree planting is increasingly understood as a way of achieving some solace in a time of grief and a way of remembering a loved one after death. Planting a tree to honor a recently departed loved one to many feels like the right thing to do. Death is difficult to deal with in part because it is a sorrowful event that offers little that one can do—planting a tree or multiple trees seems like something tangible that can be done, a tangible step that can ease the pain and grieving, as well as create something of lasting benefit to others. And the planted tree, a specific tree in a specific location can create a place of reverence, and an ongoing and permanent place of remembrance.

FIGURE 9.3 Cemeteries and Trees Often Go Together (University of Virginia Cemetery). Photo credit: Tim Beatley.

In the era of online death notices, it is now a common option to commemorate a loved one through the planting of trees: "Celebrate their life: Plant trees," says the site, with a prominent "Select a Forest" button. It is an attractive alternative to sending flowers or donating to a chosen favorite charity. This sort of planting of "memorial trees" is often done through a partnership with organizations such as the Arbor Day Foundation.[42]

A San Francisco-based company called *Better Place Forests* has taken this idea a step further. They now operate ten memorial forests around the US. It is "a natural alternative to cemeteries," declares its website, "beautiful, sustainable and full of life."[43] Here one can reserve their own tree (or a tree for an entire family) in advance of death. The tree becomes the site for mixing cremains (cremation ash) and for the end-of-life ceremony for family and relatives. Permanent access to the tree is ensured through the granting of an "irrevocable license." In addition to the choosing of a memorial tree, Better Place Forests will also plant a number of other new trees in the area (through a partnership with One Tree Planted, a company described later on).

In this model, trees and forests benefit from the normal and natural process of death. The resources and money typically devoted to a casket, concrete vault, conventional graveyard burial are then steered toward the conservation of forests. The company's memorial forest in the Berkshires of Massachusetts, resulted from the purchase of a 200-acre wooded parcel that will now be conserved and protected permanently.[44] The company's website boasts of protecting more than 1000 acres of land.

This option also helps address a concern Lee Webster of the Green Burial Council has. She expressed to me worries about the rapid growth in cremation and the fact that for many families there are often feelings of a lack of closure. Without the traditional burial service and grave to visit later, she worries about the diminishing chances for healing and grieving. Green burial options such as a memorial forest provide more of those kinds of opportunities, without the downsides of a traditional funeral. And they are in line with the perhaps unstoppable trend of cremation as the preferred death choice. Interestingly there has also developed a desire for an even more natural option for dealing with a dead body—the emergence of human composting, which is now legal in five states (and the number is rising).[45] In an era of climate change when we worry even about the energy consumption and carbon emissions of burning the dead, natural decomposition is increasingly popular. And it also aligns with memorial forests, as the resulting "compost" can just as easily be applied to the site of a memorial tree as remains can.

These options will undoubtedly grow and there is much to recommend any ability to steer at least a portion of the billions spent on conventional burials on tree- and forest-based end of life strategies and ceremonies. I can imagine a number of smaller, city-specific versions of the Better Place Forest model popping up in response to the growing demand for options.

There are many other important events in the life of a community, good and bad, that need remembering and commemorating and often the planting of trees. Such events or milestones could be national in focus—for instance the Queen's Green Canopy—a UK tree planting initiative commemorating the Platinum Jubilee. This initiative has been extended to permit the planting of trees to honor the passing of Queen Elizabeth.

In many places in the US trees and forests have been planted to honor those who lost their lives in the 9–11 attack. The 9–11 memorial in New York City incorporates 400 Swamp White Oaks around the perimeter of the two dramatic reflecting pools, with water flowing down the gaping holes where the footprints of the twin towers were. The trees soften the site adding reverential nature. The crowns of these trees, once mature, "will knit together to form a dense overstory sheltering the entire plaza, except for the voids, which will remain open to the sky."[46]

In many other cases the events commemorated with trees are very local and often traumatic—for instance to honor those killed in a local mass shooting, something all too common in US cities.[47]

Planting Trees and forests have also become important ways to commemorate important people and events in our lives. This practice could be expanded even more broadly and in ways that further solidify the emotional connection to trees and status of trees. Examples abound. In the case of my own family, there is a "Beatley" tree in a prominent park (Fort Ward) in the city in which I grew up. Visiting this tree has been, for my sister, especially a measure of solace. It is not surprising then that we engage in tree planting to remember those who have passed away.

Gratitude Gatherings

There are a variety of other ways that communities might engage trees and forests in remembering events and honoring residents. And we must also not forget the power and importance of tree rituals of various kinds. During the global pandemic visits to trees and forests for many became a daily ritual, a saving grace in many people's lives. The Icelandic Forest Service, in the beginning months of the pandemic, famously recommended that citizens seek out trees and hug them, suggesting this was a very good way to start one's day.[48] They even gave specific instructions about how to go about this including a series of photos showing the many different ways a tree could be hugged.

Hugging a tree, or touching its branches or leaves, or sitting under a tree for even a few minutes each day, can indeed be a therapeutic and healing act.

FIGURE 9.4 A gratitude gathering organized at a site in West Seattle, where two old Douglas Fir trees were soon to be cut down. Photo credit: Tim Beatley.

A daily walk to visit a favorite tree can be a collective act of reverence and thanks. Cities should help facilitate these daily tree rituals and encourage unique and creative ways to connect with and thank the trees around us.

In Seattle, the summer of 2023 saw the beginning of a series of "gratitude gatherings" organized by Seattle Tree Action and the Last 6000. They were intended as a way to raises awareness about trees that were soon to be cut down. It became a way to publicly thank these trees and to say goodbye. I had the chance to attend one of the gatherings in early August and to see first hand their power and impact. Specially I participated in a gratitude gatherings for two large Douglas fir trees at a building site in West Seattle, that were soon (the next day we understood) to be cut down. The event was attended by some forty or more people and was quite emotional. One speaker talked about the merlins (a species of falcon) that had nested each year in one of the trees, and how important the trees were ecologically. The organizers at the end of the program invited attendees to come to the front to place a rose petal or flower on a mandala lying on the ground. Many quietly walked to the trees to touch them or say something to them.

The idea of the gratitude gatherings began a few weeks earlier when a tree that became known as Luma was threatened with being cut down. This large western red cedar was located on a development site in the Wedgwood neighborhood and permits had been issued by the city allowing them to cut down the tree to make room for six market-rate housing units. During this first gratitude gathering some attendees felt it was premature to give up on saving the tree and took action, with several individuals soon sitting in the tree to protect it and raise awareness (known as "droplet" the individuals sitting in the tree wore face covering and sought to protect their identity). The media attention in the end helped to save the tree. The Snoqualmie Tribe came to believe the tree was a "culturally modified tree" and as such sought protection from the State of Washington's laws protecting archaeological sites. In the end the developer voluntarily reduced the number of housing units to five (a hollow gesture many believed as it had been shown how redesigning the development could have allowed the tree to be saved and the building of the original six units of housing).

New Ways to Radically Regrow Urban Forests

Many of the cities discussed in this book have already set ambitious tree planting and canopy targets for the future. Cities like Pittsburgh and Richmond have set relatively high goals of reaching an impressive canopy of 60%.

How cities will reach these high levels is an open question but there are many new ideas and methods that may be useful. The quick emergence of drones and use of drones for a variety of societal purposes including to plant trees and forests is one promising technology.

Lack of resources and personnel to physically plant new trees in and around cities remains a challenge, though as I have argued that tree planting and tree care really, really any form of engagement with trees, is a good thing for citizens and thus good for cities.

One intriguing idea for expanding forest in and around cities is what has been called "passive rewilding." Passive rewilding is defined by Broughton et al, as "spontaneous development of ecosystems without direct intervention, such as on abandoned land and occurs via natural succession."[49] In their study of woodland regeneration, they provide some strong evidence for the value of such approaches. They specifically studied the extent to which abandoned farm fields naturally regenerated from adjacent ancient woodlands. This happened quite successfully, it turns out, as windblown seeds were disbursed as well as seed distribution from birds and mice. They conclude from this study "that restoration and expansion of native woodland can be rapid on former farmland existing forest." While the resulting regenerated woodlands exhibited less species diversity than the ancient woodland, nevertheless after only a few decades these new forests were quite diverse.

The authors describe in more detail the natural trajectory of regeneration that takes place "as colonization of woodland shrubs and trees begins spontaneously soon after land abandonment. A rapidly developing shrub-thicket phase in the initial two decades after abandonment progresses to increasing tree cover, and eventually achieves almost total closed-canopy mixed-species woodland after approximately 50 years."

To fully employ these ideas of passive rewilding or passive regeneration in cities will require some careful planning, but there may be many places where simply leaving in place an existing old growth forest, even a small stand would serve as the seed source for the regeneration of adjacent empty or abandoned lots and fields in a city. Might we even begin to cordon off partitions of suburban lawns or otherwise landscaped areas between buildings in cities as passive regeneration sites? Just letting them alone, intentionally letting them lie fallow, might be one very effective (certainly cost-effective) approach to urban reforestation. There would need to be some steps taken to ensure the sense of intentionality is conveyed to the larger public. This might take the form of explanatory signage or frequent mowing of edges or strips to convey to the world that this is not an unkempt site.

They also found no significant negative effects from browsing animals like deer, though at least in the ancient forest in their study there has been the practice of culling herds which might help with this marginally. Whether the proliferation and growth in population of white-tailed deer in North America would doom the early regeneration of abandoned sites and jeopardize young saplings is unclear, perhaps the creative use of low-cost movable fencing and the creation of a network of small forest refugia might effectively solve this concern.

Conservation Beyond a City's Boundaries

The challenge of saving forests today will require the efforts of cities everywhere, especially the more affluent cities of the global north with more resources and wealth and more culpability for the destruction and deforestation that has happened.

Thinking beyond the narrow legal and jurisdictional boundaries of a city will be necessary and a major element of the idea of forest urbanism. This can happen in a number of ways. Cities can support and nudge the protection of forests in the larger metro area or region in which they sit. In Seattle, there have been remarkable efforts to protect forests well beyond the borders of the city. One promising tool is a regional-scale use of Transfer of Development Rights (TDR).

The role of a local conservation nonprofit, Forterra (formerly the Cascade Land Conservancy), has been essential to the success of this effort. Forterra has itself set a remarkable 100-year conservation goal of permanently protecting 1.3 million acres of prime farm and forest lands. I recently spoke with Nick Bratton, who has been involved in much of this Forterra work and has helped to design some seventeen local TDR programs as part of assembling a larger regional network of TDRs. Bratton described it as a kind of "European Union for TDR" in which participating localities agree to coordinate and allow the transfer and movement of development tights across the region. Under the program a forest owner could agree to protect land in say Pierce County, selling her rights to a developer who could use them to increase density in another jurisdiction, say Seattle, where much of the demand for density and growth exists.

The program has accelerated the supply of development rights and the use of the TDR tool, Bratton told me. There is now a regional marketplace, with all thirty-five cities in the region eligible and participating. An important part of the story has been the creation of new incentives for cities to participate in the program and this has come in the form of allowing participating cities to take advantage of new infrastructure funding through TIF—or tax increment financing.

A specific example of how the regional TDR works can be seen in the case of the Rainier tower in downtown Seattle. Through a TDR transfer the developer of this tower was able to add an additional 200 units to the project. The purchase of these units results in the permanent conservation of 14,000 acres of forestland. In this way the benefits flow both to the developer (adding helpful density where it is appropriate and needed) and for owners of forestland, selling their development rights can be a very attractive option as well.

Partly this is about rethinking the role of cities in the conservation of larger, global ecosystems. Can and should cities think about how they might support global forest conservation, protection, and conservation of forests many hundreds or thousands of miles away?

A project of the World Resources Institute (WRI), "Cities 4 Forests" has been helping cities around the world to think beyond the forests within their boundaries, already with some considerable success. *Cities4Forests* is a relatively new network of cities, what program manager James Anderson calls a "global alliance of cities," now numbering ninety-one, who agree to work toward conservation of forests.[50] As Anderson notes, "Many cities don't have a good starting point with which to engage in a conversation about forest conservation that far away from their city walls."

The project thinks in terms of three scales and the need for action by cities on all of them. There is not only the local scale—the many things' cities can do to protect and restore forests within their legal boundaries—but also "nearby forests" and "faraway forests." To join, cities must endorse a political statement on the global value of forests and commit to engaging on at least one of these scales.

With staffing and expertise from WRI, there are now a number of positive examples of how the project has helped cities make progress on these three scales. At the scale of "nearby forest," there has been considerable success and a number of positive examples to point to. These include the example of Little Rock, Arkansas, and their work with the local water utility to protect land around the main water supply reservoir. Specifically, the utility has employed a green bond to raise some $31 million in funds to support this initiative, with $7 million used to purchase 4300 acres of forestland.

Several international examples include work with the cities of Jakarta (Indonesia), Xalapa (Mexico), and Vitória (Brazil). In the case of Vitória, WRI helped the city develop a program to restore upstream watersheds degraded from many years of cattle grazing. In Xalapa, an innovative voluntary water fee was used to fund, again, to protect and restore upstream watersheds and in this case to overcome a lack of potable water supply for thousands of residents. In Jakarta, WRI has helped support the city in the development of new regulations to protect existing trees and forests, as well as to follow through with its commitment to plant 200,000 new trees.

Much of the WRI work around Cities4Forests is focused on helping cities develop new ways to understand and calculate the economic benefits of forests, and especially for cities in the Global South to develop baseline data on forests (such as measurement of a city's forest cover) and to develop forest maps and mapping systems that are commonly in use in more affluent cities in the Global North. Developing financial tools and strategies seem another key priority, building on the special staff and resources of WRI. One recent example of WRI's work helping to develop new conservation funding for forests can be seen in the promising idea of Forest Resilience Bonds. A direct response to the growing economic and ecological impacts of wildlife, especially in the western US, is that beneficiaries of better forest management (e.g., selective thinning and controlled burns) help to pay for these new management demands. An initial testing of this idea has raised $25 million to better manage 45 million hectares of forested lands around Lake Tahoe.

How cities can take actions and make a meaningful difference in conserving faraway forests is a greater challenge still. One focus of the work of *Cities4Forests* has been to help cities understand the global impact of their consumption patterns. Urban demand for beef and agricultural products, as well as timber, are major drivers of global deforestation, of course. "Cities are the major center of consumption for these commodities," says James Anderson, "but very few have a sense of exactly what their forest footprints, their deforestation footprints, and resulting climate footprints [are]." To address this, WRI is developing and will soon launch a "platform and methodology" to help cities better estimate these impacts as well as to develop strategies to reduce them.

Cities in this network can also engage in "political action," Anderson notes, and can provide a strong "voice" for global conservation. He mentions how the network orchestrated the signing and issuing of a political declaration, a "Call to Action on Forests and Climate," signed by more than fifty mayors.

Yet another set of approaches has to do with significantly reducing the ecological footprints of cities and the long-distance impacts on forests of our lifestyles especially in opulent and high consumptive cities of the Global North. There are several options. Cities have a number of policy levers they can deploy in an effort to reduce the consumption, for example of illegally harvested tropical hardwoods. Some years ago, New York City adopted a policy aimed at discouraging the purchase of tropical hardwoods. Municipal procurement continues to represent an important mechanism especially in larger cities with immense budgets. There has been a gradual growth in the development of green procurement, with many cities now imposing sustainable requirements for the many products and services they purchase.

Cities can also begin to take steps to conserve and protect forests beyond their boundaries by recognizing the larger benefits and services provided by the larger watersheds and ecosystems in which these cities are embedded. New York City, again, has been a leader and one of the best examples of this can be seen in that now fairly well-known example of the city buying and protecting land in the Catskills as an essential step in protecting the main source of potable water for the city.

Cities can work together to conserve these larger ecosystems that extend beyond any single city border. In the San Francisco Bay Area, for example, a unique and successful approach is being taken to conserve and reswide parcel tax. Perhaps similartore the Bay's wetlands, through a unique bay-wide parcel tax. Perhaps similar approaches could be taken to protect regional and bioregional forests that provide nearby cities with important ecological services.

What about the role of cities in the conservation of forests that may be thousands of miles away? Here, emotional connection and connectedness with these forests, and physical and ecological relevance will be harder to establish. But especially larger, richer cities in the Global North have opportunities and the resources to take steps to protect these more remote forests as well.

Reducing the forest-impacting ecological footprints of these cities is a major challenge. We know that the highly consumptive lifestyles of Northern cities fuel the conversion of tropical rainforest to agriculture (including palm plantations) and to cattle ranching (to satisfy demand for beef).

Many cities have significant global investment funds, for example the large pension funds for teachers managed by New York City. Divesting from fossil fuel companies has become a priority, and such shifts are helpful in conserving forests as well. Cities might also seek to influence more indirectly the forest-destructive consumption patterns of its residents—for instance, through promoting the idea of meatless Fridays or other ways that discourage forest-impacting or forest-consumptive behaviors.

But cities could do more and could go further. Spending city resources directly on global forest conservation programs and initiatives ought also to be on the table. This could be direct funding for grassroots forest conservation organizations operating in other countries or tree-conservation and tree planting connected to offsets for a city's carbon emissions.

In September 2022, the Norwegian and Indonesian governments announced a new agreement in which Norway would pay Indonesia for reducing its levels of deforestation, and thus its emissions of carbon. Might it make sense, in an era of growing city diplomacy, for cities like Oslo and London, to enter into similar agreements with cities in the Global South, perhaps, that would compensate for reaching forest conservation milestones.

Carbon Offsets to Protect and Grow Forests

I was introduced to the concept of environmental offsets and its potential twenty years ago when planning to attend the UN Conference on Sustainable Development in Johannesburg, South Africa. Delegates and attendees were encouraged to purchase "offsets" that would counterbalance the carbon emissions connected to the airline travel needed to get to the conference.

In my case I purchased offsets from a nonprofit called Future Forests that agreed to plant ten trees to compensate for the impact of my travel. I later received a certificate that told me the location of the tree planting and provided a map of where in Mexico (two sites in the state of Chiapas) they were planted. That experience illustrated for me not only the advantages but also the limitations of offsets. It was relatively easy to make and process the payment, and to take direct responsibility for the carbon I was causing to be emitted. Yet I also (to this day) do not know for sure that the trees were actually planted or that they have survived to provide the environmental benefits promised.

The potential is great indeed for citizens of cities around the world, but especially in cities in the more affluent Global North, to harness their travel in these kinds of ways for forest conservation. As travel picks up as the global pandemic lessons, citizens of cities like New York, could do much to stimulate tree planting and forest conservation. In the summer months of 2022, for instance, more than four million New York City residents travel abroad each month. If only a quarter of these travelers planted trees, or paid organizations to plant trees, this would still amount to a massive tree planting or forest conservation efforts.

Offsets have become controversial in recent years and there is considerable evidence that they often do not provide the benefits promised.[51] The recent case of the Massachusetts Audubon Society selling carbon credits for protection of forested land it owns has received some critical and unwelcome attention.[52] The issue in this case is a common concern with offsets—is it true that the purchase (say by a company or other carbon polluter) will result in protecting forests that would otherwise be lost? This is often described as the problem of "additionality." It was not at all clear for instance in the Audubon forest case that the offsets resulted in real forest conservation (would the forest have been lost without the offsets, perhaps not in this case, so little additionality likely resulted).

Yet, offsets have been effectively used in settings and circumstances closer to home. Austin, Texas, for instance, has set the goal of being carbon neutral and taking many steps to move in that direction. Part of its efforts have included buying carbon credits from tree planting projects in the greater Austin region. A local nonprofit called TreeFolks planted the trees in areas outside the city, in part to reforest floodplains, with the carbon credits verified by Seattle-based nonprofit City Forest Credits.[53] In this way, Austin has helped to fund the planting of thousands of new trees, restoring riparian habitat and sequestering carbon, and helping it move toward its goal of carbon neutrality.

Linking with a local nonprofit like TreeFolks is perhaps a better approach than many, and focusing on offsets that lead to tangible forest outcomes not too distant from the city buying them (they can be seen and visually verified to make a difference) is probably a good idea. Offsets when done in these ways could lead to significant positive results.

Another way to understand the impacts of global travel is to recognize the immense import of what visitors from other cities and regions see and experience when they visit cities like New York. In New York City nearly 67 million visitors came to the city in 2019. The power of a positive example can be seen in the example of the High Line. While not a perfect example, the park has become a popular destination for visits and help to show how underutilized or even abandoned infrastructure in cities could be the basis of new recreation, respite, and nature in cities.

It is also possible for cities to develop programs and initiatives that seek to directly connect with and to support and influence forest conservation in more distant locations. Cities can attempt to steer their often-considerable spending power in ways that support sustainable harvesting of timber and wood, and in this way significantly reduce the impact of the large ecological footprints of cities.

One of the most creative and promising initiatives in this regard is the Brooklyn Bridge Forest. The winner of the Val Allen Award, and a project supported and profiled by Cities4Forests, it seeks to forge a direct relationship with a distant forest from which timber will be sustainably sourced. The forest referred to in the Brooklyn Bridge Forest is not a *local* forest at all, but a tropical forest located far away in the highlands of Central America. In this creative scheme this distant forest becomes the source of the 11,000 hardwood slats needed to refurbish the iconic Brooklyn Bridge pedestrian walkway. It will be a "landmark forest protection model, generating wood for the Brooklyn Bridge Promenade in perpetuity," say the project materials, and leading to a "first-of its kind partnership between a major city and a forest of global ecological and cultural significance."[54]

The project will notably "link people across geographic and cultural boundaries in support of environmental protection and historic protection."[55] More specifically this "partner forest," in this case is the Uaxactun Community Forest, located in Guatemala in the Maya Biosphere Reserve. As a result of the economic support provided by New York City (through the purchase of wood) will in turn lead to the protection of an estimated 200,00 acres of forest. New York residents will also have the chance to participate in the conservation project directly, through sponsoring individual planks on the bridge. "Each board sponsored would protect the equivalent of 18 acres of forest."[56]

It remains a serious challenge to imagine how the vision of "forest urbanism" can truly assist in the conservation and protection of global forests. We should be open to any discussions about how cities can reduce their ecological footprints, in general and specifically connected to forests. We know that the diet of urban residents has a huge impact and so perhaps that is an important point of leverage in the future.

Already cities exercise some influence over diet and can help in fostering more forest-friendly diets, specifically that rely on a lesser amount of meat and meat sourced from more sustainable sources.

There are official "Meatless Fridays" in many cities, and efforts to reduce meat consumption in city schools. Perhaps even more radically, cities and urban population might strongly support the further development of so-called cellular agriculture, finding long-term and more sustainable alternatives satisfying the beef and protein needs of the larger world. The company called Good Meat has been a pioneer in developing this market, and Singapore is the first country (or city-state) to have made the sale of meat produced through these cellular-ag technologies legal. Here's the pitch on their website:

GOOD Meat is real meat, made without tearing down a forest or taking a life. We are the first and only company in the world to sell cultivated meat. Our first product is GOOD Meat cultivated chicken, which Singapore has approved for sale, and is currently available at select restaurants."[57]

Are people ready for cultivated meat? It is not clear, though many of the regulatory hurdles seem to be being overcome. Will consumers overcome their squeamishness about it? This is not clear. There is something unattractive about the food we eat being produced in bioreactors, though this narrative could change. Uma Valeti, founder of a company called Upside Foods, says this that may be reassuring: "The process of making cultivated meat is similar to brewing beer, but instead of growing yeast or microbes, we grow animal cells."[58] And it is "Meat without slaughter," something many consumers will be attracted to.

A New Ethic of Tree Conservation for Cities

What we are in need of is a new ethic of trees and forests, especially in cities, an ethic that would apply at not only collective or civic levels but also at the level of an individual. The current way we see trees in cities and care for those trees is inadequate and at the heart of many of the conflicts we see play out in cities. For many, perhaps especially those managing trees at a distance, trees are seen as little more than a form of street furniture, no more important and largely interchangeable with a bench or a plastic play structure in a park. Rarely, it seems, are they viewed as living things and as something providing essential services, highly valuable and in the case of larger older trees as irreplaceable, at least in any short-term timeframe.

Part of the problem of course is that we have a hard time seeing these trees in the first place. Even if we are lucky enough to live in neighborhoods with abundant trees, we often do not know much about them. Kids and adults alike exhibit a shockingly low level of tree literacy; we are lucky if we are able to identify even a handful of local or native species of trees.

The general unimportance of trees is demonstrated daily in the lack of care we give them. This was recently demonstrated in a visceral way for me, at my own home, when we had to contract with a plumbing company to replace a leaking water pipe, connecting our home to the city's main line at the street. I arrived on this day to find that the company, using a device that dug and inserted a new line underground, had placed a stabilizing chain tightly around our large tulip poplar. I immediately expressed my distress as this chain was digging into the bark of this precious tree. The contractor seemed genuinely perplexed that this would be a concern or a problem. Protecting the health of this tree was infinitely more important to us than how or when the water line was replaced (obviously water supply is important), but it seemed the tree was just "there," something to be used in this case to anchor a machine, with little regard for its health.

Mostly this seems to be our view of trees, especially those in cities. The sense of seeing trees as merely another form of street furniture then translates, indeed empowers the thinking that frequently surrounds the easy replacement of an older tree with a sapling or young tree—there is nothing distinctive about that older tree or about its longevity or ability to have survived and sustained over a long period of time—it is easily fungible or tradable, perhaps for another young tree or group of young trees but perhaps for an air conditioning unit or an investment in reflective pavement (which might be a good thing to invest in but should not be viewed as a substantiate for actual, living trees in our neighborhoods).

Trees in cities are often primarily understood as private property. In the real estate and the development community they might be understood as amenities that add value to a new neighborhood or development project, but they are often seen as obstacles to building, simply part of the background conditions of a site, and a focus of site preparation—clearing the land, creating a condition of terra nova so that homes and buildings can be constructed. They are the important things after all, it is felt, not the environment or the trees.

A recent example from my home city of Charlottesville illustrates this perspective well. A large area of forested land had remained undeveloped

FIGURE 9.5 The Forest at Azalea Springs. Photo credit: Tim Beatley.

for a number of years, largely because a significant portion of it (about 10%) fell within the definition of the city's critical slides ordinance (see Figure 9.5). To build there would require a waiver from this ordinance. In January 2023, the developer of this project, called Azalea Springs, sought from the city such a waiver, and in a 3-2 vote received it, over the strong objections of the environmental community and the city's Tree Commission. The developer has agreed to save a few more trees and to create a small public park in exchange for the waiver, but very few trees in this forested parcel will ultimately be preserved. It will essentially result in the near-complete deforestation of this parcel of land.

What will replace this forest will be forty-seven single family homes, presumably on completely cleared and biologically sterile suburban-style parcels. Only two of the homes will be considered affordable. In justifying his vote in favor, Mayor Lloyd Snook declared the city really had no choice—the developer had owned the land for some time and the only option the city had would be to purchase the land as a park.

"It was not an option for us to say 'No. No building at all, preserve all these trees," Snook said. "I would love it if we could. The only way we could would be [to] buy that property and preserve it as a park."[59]

There were in fact many options the city could have pursued, and had the trees been considered public assets these options might have been seriously considered. The city lacked a strong protection code, so from the start the trees had little status other than as private property. Even the Tree Commission was not advocating for total protection of the forest, though purchasing the land might have been a good adoption. Rather, as we know from other examples (described in this book and elsewhere) it is possible to develop, perhaps even at a higher density, while saving many trees. The transfer of development rights (TDR) is another option, a power that localities in Virginia have been given. A code such as Portland's would allow the forest to be preserved and the density transferred to another less ecologically damaging site, or the rights sold to another developer. There were in this case, and almost always are, alternatives. But if you start from the premise that the trees have no legal or moral status the result is almost always the careless loss of trees.

The kind of civic ethic of trees and urban forests is one that if it could take hold could guide and steer the psychology of a community in the direction of protecting trees and expanding canopies. I often wonder where the public (or private) indignation is when large old trees especially are cut down in very public places and ways. The homeowner or builder or developer seems to feel no obvious guilt or shame about the act. There are no headlines in the local paper (partly because local papers in many places no longer exist, another problem to address).

If we fear that our neighbors and larger community will disapprove of how we treat trees, partly because it is viewed as a form of collective vandalism, then there will be a strong social and psychological disincentive in place to harm those trees in the first place. Public discussions about, and ultimately an embrace of, a collective ethic that holds that trees are, to a considerable degree, civic and collective resources, will do much to protect our trees and expand our forest canopies.

We also spend remarkably little public funds to plant, grow and sustain the trees and urban forests around us, even though as I have argued here they deliver such an impressive array of public health benefits, especially mental health benefits.

In the US legal system, support for this expanded ethic of trees and forests might derive in part from the public trust doctrine, a common law doctrine dating back to Roman law, and that recognizes public rights in the use of some kinds of natural resources. The key idea is that there are certain elements of the natural environment—navigable waters and coastal shorelines, for example—that are so imbued with publicness and public use values that the public cannot be excluded from them in the exercise of private property rights. "The public trust doctrine requires the sovereign, or state, to hold in trust designated resources for the benefit of the people."[60]

A property owner along the coast cannot legally erect a fence that impedes the ability of the public to walk along the wet beach (and in several states the dry beach as well, up the first line of vegetation). The public trust doctrine, then, serves to place significant constraints or limits on the use of private property. Protection of coastal wetlands, and restrictions on their filling and destruction, are further supported often by reference to the public trust doctrine.

Urban trees and forests might in a similar way be protected under this established common law principle, and the public trust doctrine might serve to reinforce or underpin the new ethic of trees that I am arguing that we need. Should large diameter, older trees in cities be understood as natural elements that are held in trust and should be protected by the sovereign (i.e. the state or the city?) It does seem to me the level of publicness with which they are imbued makes a strong case.

Can it be said that the government has a responsibility to hold urban forests in trust (for current and future residents of a city)? Urban trees, to the extent that they are situated on private land, might be seen as something in a different category compared with a lake or coastline or a navigable river. Yet despite the reality of large trees literally growing out of the ground and deep root structures that penetrate and interweave with land and soil, they are again living systems that are so imbued with public values and benefits that the idea of managing and protecting them in trust makes considerable sense (to this author).

Rob McDonald, of the Nature Conservancy, has rightly pointed out the amazing gulf between what we spend each year on health care in the US—about $3 trillion, or around $9500 per person—and our meager spending on trees of $5.83 per tree, per year.[61] Funding for trees will pay back remarkable dividends and is a small investment in a larger public health return, yet we tend not to see it this way. Some health care companies have supported investments in trees and other forms of green infrastructure, but this is still fairly uncommon. There may be little incentive without a single payer health care system, as the improvements in public health are not necessarily reaped by any single health care provider.

We need to evolve our individual ethics to incorporate concerns for the larger public, and an appreciation that any decision to cut down a tree, or to save it, likely on our own private land, will be infused with civic and collective dimensions—that decision, especially when multiplied by thousands of other similar individual decisions—will have significant impacts on the quality of life in our neighborhoods and in our communities, as well as serious implications for public health. Saving that large oak tree cools the larger spaces beyond our own homes and together with other trees makes up a collective canopy that cleans the air, retains stormwater, cools the city, and creates habitat for birds and other creatures we all enjoy and benefit from.

This expanded tree ethic extends as well to the many other things we might do to support the larger ecosystems and environment in which these trees live. The ethic implies a duty to care for the trees in our immediate surroundings, to steward over them, to water them and care for them to ensure that they survive and thrive.

Such a broader, deeper ethic of trees will serve as a counterbalance to the prevailing or dominant view of trees as essentially private property, free to do with what one may, unencumbered by any larger set of considerations. It is this narrow notion of trees as private property that fuels a sense of indifference or casualness on the part of citizens and property owners.

An expanded notion of urban citizenship, moreover, suggests that individuals have other collective duties to the larger public. Citizenship has been construed too narrowly in the past—one is expected to obey the law, to vote at election time, to pay taxes that are owed, and when asked, to serve occasionally on a jury. It is a minimalist notion, not asking very much of individuals. Voting is already something very difficult for many of us, and voter turnout, especially for local elections, is depressingly low.

A Tree Citizen is one who works to grow and protect a healthy urban forest, and to commit to basic political actions in support of trees and other aspects of the local environment. This includes, where possible, participation in tree planting events when they occur and in support for organizations (political, economic, and moral) in the city that work to grow trees and forests. As we have seen, there are a growing number of such local organizations, but they are largely small, fledgling groups, typically under-funded.

Being ready to stand up for the existing trees in a city is another part of this tree citizenship. As the stories told in earlier chapters have hopefully shown, even strong tree protection codes require active citizen engagement and monitoring; they require local citizens to vigilantly watch over trees and forests, especially those nearby that we know the best. Actions to stand up for trees may require us to show up at a public hearing or a city council meeting, to sign a petition, to lend our voices to those who question projects that, while well-meaning, whittle away at the canopy of a city.

A better understanding of the psychology by which trees are cut is also needed. Personal experience suggests that there are a number of reasons that may motivate a property owner to cut down trees. These range from worries about the danger of trees falling down, to often aesthetics of trees, and a dislike for raking leaves in the fall, or desire to let sunlight into a home or to open up the sightlines from the street. Sometimes homeowners are motivated by misguided ideas about what will improve or make their homes more valuable (cutting trees down will likely have the opposite effect). The reasons are often trivial and frivolous, and the reasons do not really have to be very good ones if the strong starting point is that trees are only or primarily private property.

A tree ethic would hopefully serve as a constraint on these decisions and would help to create the context in which one might incur the wrath of neighbors if a tree were cut down. Much of the work of a tree code could be done through strong collective mores about trees. Aldo Leopold discussed this in his semifinal book A *Sand County Almanac* some seventy years ago. Leopold said this about the way a land ethic works: "The mechanism of operation [of the land ethic] is the same for any ethic: social approbation for right actions; social disapproval for wrong actions."[62] There is power in the informal forms of enforcement that result in (and from) a neighborhood or a community that cares about trees and conveys that clearly and strongly to fellow residents faced with choices about protecting (or not) the trees on their property.

I am reminded of the moving poem by Mary Oliver, "The Black Walnut Tree," in which the narrator and her mother grapple with whether to cut down the tree to help pay expenses. In the end Oliver writes: "Something brighter than money moves in our blood," and the decision is made to save the tree. Selling the tree might pay off the mortgage, writes Oliver, but if they did this, "what my mother and I both know is that we'd crawl with shame in the emptiness we'd made in our own and our fathers' backyard."[63]

Today there is rarely if ever a sense of shame experienced or expressed by those who have cut down trees. It is a matter of personal choice, about changing the look and feel of a home, making room for something more valuable or preferable. There is a little in the way of an overarching ethic or set of principles that would serve as a constraint on such actions; little fear that one will be held responsible or held in a collective contempt because of the cutting down of a tree or even a forest.

Cities can themselves help to underwrite and bolster this expanded tree ethic and this expanded notion of citizenship. This can happen in a number of ways, many already mentioned in this book. Enacting stronger tree codes that establish the legal (and political) protections for trees will send a strong signal about their value and priority, and they must lay out clear and rigorous criteria for limited circumstances when trees may be cut down. As well, creating economic incentives to preserve trees (and significant economic costs when they are cut down or damaged), is partly about a community sending the signals that such trees, especially older trees, are important.

Investing in effective mechanisms for enforcement and implementation of tree codes is another important step. Cities can also help support the grassroots organizations that stand up for trees. The more engaged a citizenry is with trees and forests the stronger will be the local democracy overall.

A tree ethic would of course be underpinned by the recognition of biophilia, the innate affiliation humans have with nature, the result of a long period of coevolution. Stephen Kellert says "The moral imperative of biophilia is that we cannot flourish as individuals or as a species absent a benign and benevolent relationship to the world beyond ourselves of which we are apart."[64]

The Rights of Trees and Forests?

We must begin to see trees as part of a larger community forest. Many of the shortsighted decisions to cut down large older trees described in this book are at least partly the result of a strong sense of the private nature of such decisions—we are told (and reinforced by the legal system) that trees are part and parcel of our own private property; we own them and the land around them and it is our right to decide when and if to cut them down.

What we need to a large degree is an expansion of the perspective of trees, a shifting from the overly narrow view of them as private property and an explicit recognition of their publicness; that is, that your decision to cut down that tree has an impact on others and on the larger public. It is not just your individual decision but the similar decisions of others who live around and in, and benefit collectively from the trees and forest.

In short, we need to begin to see our individual trees or our backyard small grouping of trees as intimately connected to a larger whole, as part of a larger community forest. If we see trees in this way there is a greater chance we will reconsider that decision to cut down a tree in favor of protecting it. How we cultivate a sense of being part of a larger community forest is an open question, but many believe part of the answer must be attaching new legal protections or rights to these trees.

There are many potential ways to create legal protections for trees and forests, of course. I have already discussed the many different approaches to local tree codes and ordinances, I do believe every city should adopt at least minimum protections through this means.

Private citizens could take steps to stipulate protections for beloved trees or groves, though this seems not to happen very often. When homes or property are sold typically few conditions relative to trees are placed on the future buyer. Trees are free to be cut down as the new owner wishes and there is no assuarnce an anicent tree will be protected and cared for. A prominent, example from Charlottesville (Virginia) to the contrary can be seen in the actions taken to proetct the so-called Earlysville White Oak, the second largest white oak in the state and likely over three hundred years old. When the church that owned the land on which the oak sits sold it to the city's airport, they insisted the tree be protected.[65] Saving trees in this way could become more common, and might happen through inclusion in a will, or through the placement of a tree preservation covenant that runs with the title to the property, or a tree protection easement held by a third-party. Perhaps an urban homeowner or landowner ought to be encouraged to award legal title to the tree to itself! One rather famous example of this, and an early application of the concept of the rights of nature, is the story of the tree, in Athens, Georgia, "that owns itself." The owner of this tree, Colonel William H. Jackson, famously bequeathed in his will the ownership of the tree to itself, including land within eight feet of the tree.

The following words on a plaque can be found on the site:

FOR AND IN CONSIDERATION
OF THE GREAT LOVE I BEAR
THIS TREE AND THE GREAT DESIRE
I HAVE FOR ITS PROTECTION
FOR ALL TIME, I CONVEY ENTIRE
POSSESSION OF ITSELF AND
ALL LAND WITHIN EIGHT FEET
OF THE TREE ON ALL SIDES
                              - WILLIAM H. JACKSON (c. 1832)[66]

The original tree was thought to be a 400-year-old White Oak and came down in a storm in 1942. Members of the Junior Garden Club of Athens took it upon themselves to plant a replacement oak grown from an acorn collected from the same location. The Junior Garden club have become long-term stewards of the new tree.

More about the role of the Club can be found on its "history" page:

The Junior Ladies' first permanent project began in 1946 when a seedling propagated from an acorn of the Tree That Owns Itself was planted at the site where the original aged white oak fell on the ground deeded to the tree. For fifty years the club maintained the historic tree site. In 1996 a community birthday celebration was held with local officials and Junior Ladies' original president in attendance. Special guests included the members of the Athens Garden Club Council, organized by the Junior Ladies in 1951, and "favors" were seedlings of acorns from the famous tree.[67]

The broader subject of the rights of nature has been growing in importance and has shifted from simply a set of ideas or theories to being put into practice in a number of countries around the globe. It can accurately be described as a movement and has been gaining ground even in the US. The ideas and practice have special significance for urban trees and forests.

I first became acquainted with the idea of the rights of nature through an essay by law professor Christopher Stone, later published as a book, *Should Trees Have Standing? Toward legal Rights for Natural Objects.*[68] I was exposed to this in an economics of property class at the University of Oregon in the early 1980s. To me it was a refreshing idea but one that seemed unlikely to find an embrace at least in the western legal system.

Stone had a long career in legal academia, teaching for half a century at the University of Southern California. Stone grew up in Washington, DC, and when he died at the age of 83, his obituary mentioned the importance of time spent exploring the nearby Rock Creek Park, "where he liked to collect turtles."[69] His obituary tells the story of Stone's insistence during the design of an addition to their southern California home that a loquat tree, from which he made jam, be saved.

Stone begins his argument by noting that the idea of assigning legal personhood to a lake or river or forest is not as odd or unusual as it might seem at first. "The world of the lawyer is peopled with inanimate right-holders: trusts, corporations, joint ventures, municipalities ... and nation-states, to mention just a few." My own employer, a large public university, is itself like a corporation, a rights-holder and acts like a legal person or entity. It is able to institute legal actions, and it is also able to press the courts for legal remedy and financial awards when, say, a contract is violated or its interests are harmed.

This is of course not the usual way we treat a river or a forest. When the Exxon Valdez oil spill released 11 million gallons of oil, massively killing wildlife and damaging sensitive ecosystems, it was not the fish or sea otters whose interests were taken into account but the economic and other effects on human beings and human livelihoods that were calculated and that ultimately became the basis for court-awarded damages.

Thus, for an urban tree or a forest to rise to the level of a corporation or a municipality even, the legal system must begin to treat these living things differently. For trees and forests to truly be "rights holders," Stone points to three main criteria that must be satisfied:

"first, that the thing can institute legal actions *at its behest*; second, that in determining the granting of legal relief, the court must take *injury to it* into account; and, third, that relief must run to the *benefit of it.*"[70] (Emphasis in original)

Stone points out that there is a long history of the expansion of rights over time, and that assignment of rights to many categories of humans occurred gradually.

Aside from the actual or specific legal protections that assignment of legal personhood might provide, perhaps the larger effect or impact is the official (political and legal) acknowledgment that nature—a river, a forest, or a bird—is worthy of protection and respect. It puts them in a different category, pulls them into an expanded definition of our moral community. "Until the rightless thing receives its rights, we cannot see it as anything but a thing for the use of 'us', he wrote. It is as much a change in the perception of nature—a strong collective statement that nature and natural elements are intrinsically valuable and important and that humans are duty-bound to respect them.

There has been remarkable progress around the world in embracing and applying the idea of the rights of nature. Notably we see in New Zealand the assignment of rights of personhood to a river, and a forest, and in Ecuador, the constitution has even been amended to give nature legal rights (with strong legal opinion in the Los Cedros case, as evidence the ideas are decidedly taking hold there[71]). In 2021, the Mutehekau Shipu River (also known as the Magpie) was granted legal personhood in Quebec, Canada, with specific rights acknowledged (Table 9.1).

In September 2022, the Spanish Parliament granted legal rights to the Mar Menor, an enclosed saltwater lagoon, the first example of such an action in Europe.[72] Here, water quality and biodiversity have suffered as a result largely of agricultural runoff. Spanish law creates a framework in which a scientific committee will serve as a guardian for the lagoon.

Despite the positive and promising applications of these ideas in New Zealand, Ecuador, Canada, and elsewhere, it has encountered a more difficult path in the US. The history and emphasis placed on individual (human) property rights, and an especially virulent form of individualism,

TABLE 9.1 Nine Rights of the Magpie River

1 The right to flow;
2 The right to respect for its cycles;
3 The right for its natural evolution to be protected and preserved;
4 The right to main its natural biodiversity;
5 The right to fulfill its essential functions within its ecosystem;
6 The right to maintain its integrity;
7 The right to be safe from pollution;
8 The right to regenerate and be restored;
9 The right to sue.

Source: Summarized in Stuart-Ulin, 2021.[73]

in combination with power held by corporations and the private sector, have made it a harder sell here.

But there have been inroads and successes in some cities. Pittsburgh, a member of the Biophilic Cities Network, took the impressive step of adopting a code that establishes a strong statement of the rights of nature. The rights of nature ordinance passed the city council unanimously.

More specifically …

(b) *Rights of Natural Communities.* Natural communities and ecosystems, including, but not limited to, wetlands, streams, rivers, aquifers, and other water systems, possess inalienable and fundamental rights to exist and flourish within the City of Pittsburgh. Residents of the City shall possess legal standing to enforce those rights on behalf of those natural communities and ecosystems.

The Pittsburgh code and statement were very much motivated by a specific concern about fracking, a practice that has been legal and highly damaging throughout much of the state of Pennsylvania. Referencing the rights of nature statement, the city chose to ban commercial fracking within its city boundaries. While affecting a relatively small area, in a city, it was nonetheless a bold move.

A case in point is the story of efforts to establish and implement a Lake Erie Bill of Rights (or LEBOR for short). Largely a response to the failures to protect the lake and its water quality from the damaging runoff and pollution from agricultural operations around it, the citizen-initiated campaign led a referendum and public vote, with a resounding positive outcome—with LEBOR later challenged in the courts and found not to be constitutional, a setback for supporters to be sure. The campaign also served to highlight and foreshadow the kind of opposition likely from farming and other vested interests who see the rights of nature as a threat. During the campaign, some $300,000 of outside money was spent in advertising against the measure, and at the time the source of these funds was unknown. Later it was discovered that the anti-LEBOR funding came from a Houston, Texas, foundation connected with the oil and gas industry. I am not optimistic that corporate interests and extractive industries will sit by idly and watch the expansion of legal rights for forests, rivers and mountains. As in the LEBOR example, corporate and extractive industries will hit back, and able to use the immense economic resources they have at their disposal.

How as a practical and legal matter the rights of a tree or a forest get protected is unclear. One of the main requirements for this idea to be truly put into practice is to figure out who (what person, institution, agency, organization) would have the ability and duty to stand in for the tree or forest. Stone imagines some system for assigning or designating "guardians," something he notes already exists in states like California where the court can designate a legal guardian in cases where a person is incompetent or unable to take care of themselves or their affairs.

Who stands up for trees and forests remains a critical question. What institution or organization is to be given the duty to give voice to their interest? And what specific powers will this body or organization have and will be sufficient? In the Azalea Springs example from earlier in this chapter, arguably our Charlottesville Tree Commission served in a capacity as a guardian voice, yet ultimately their views were dismissed and lacked the actual power and authority to do anything meaningful to protect the forest.

I spoke recently to Dutch environmental lawyer Jessica den Outer, who is writing a book on the rights of nature, and who believes in the need to shift our thinking and our law in this direction.[74] She describes her own legal education and not being exposed at all to the concept of rights of nature, until she graduated, and heard about the now-remarkable court decision in Ecuador so-called Los Cedros decision. In it, the Ecuadorian supreme court found that permitting mining companies to destroy the cloud forests there amounted to a violation of the rights of nature.

Den Outer is optimistic about the eventual shift in this direction, though she agreed with me that it will probably take a more fundamental shift in the thinking and support of the larger public. "People do change their views," she told me. She sees her mission now as largely about educating the public, which we both agreed will be essential to ensuring that legal opinions like the Los Cedros case, when they happen, are enforced.

She recently participated in an interesting event in Amstelpark, in Amsterdam, largely aimed at this kind of broader public education. Part art project, part legal debate, she and another lawyer argued the case for the rights of the trees in the park and elsewhere in the Netherlands. The October 2022 event was called "Lawsuit in the park—the trees speak their mind." The photos of the lawyers in barrister garb standing in the midst of a grove of old trees is striking and was likely curious to those who didn't know what was happening. Den Outer and her legal colleague argued on that day not only for rights of those trees around her but also for trees in other cities. She explained that it was common practice for cities there to cut down trees after they reached the age of thirty or forty years old. It was a dramatic event, and I wish I could have been there to participate and see it unfold. As she explained, it was also a demonstration of how law can combine with science and art. A scientist explained how much carbon the trees sequestered while artists and musicians showed how to represent these rights and how to stimulate thinking about them. The project and event received a fair amount of media attention apparently.

Den Outer told me of a group of young people she has been working with who are devising a scheme for how a forest near the Dutch city Doorn can "own itself." The Doorn forest is relatively small, around six hectares in size, or about fifteen acres, but a good size for a test case, and there is an even greater forested area as part of the larger estate in which it sits. "They're trying to give it back to itself," she explained. She serves on the supervisory panel for this intriguing and promising project.

I reached out to the founder of the initiative, Peter Akkerman, whose day job is working on climate and diplomacy issues for the national government in The Hague.[75] He explained that the idea was gaining traction, and that in addition to the forest in Doorn there were several other forests they were in conversation with about similarly establishing and testing the idea of forests that own themselves.

To advance these ideas, Ackerman has created an unusual sounding foundation—"Stichting Bos dat van zichzelf is," in Dutch, or "the forest foundation that is its own."[76] Ackerman is finding himself spending much of his spare time speaking about the idea and spreading the word. Not all of the practical issues of establishing such forests are worked out, and he and I discussed a number of uncertainties. Who or what would have the power to stand up for these forests and to represent them? And how will it be possible to understand and identify the best interests of the forests? He explained that they have recently been running a series of workshops designed to help people "listen" to trees and forests, and some of the answers to these questions will come in this way.

I asked about skeptical opinions he might have encountered about the idea of forests owning themselves. What will happen, for instance, when there are perhaps many small forests in a region that have been declared forests that own themselves? Will that mean it becomes extra difficult to build roads or expand a high-speed rail line or to undertake other collective projects? That though seems the point to some degree—it should become more difficult, for instance, to build an additional runway at Schiphol airport, if to do so results in the loss of forests.

Another practical question has to do with the level of active management, if any, that will be permitted in these forests under this idea. Akkerman explained that for now he and his group are imagining that forests would mostly be left alone, but conceded there could be times in which humans can and should intervene—for instance to control an invasive pest or to help the forest to become more biodiverse (many existing forests in the Netherlands, Akkerman explained, have been planted as plantations and consequently are not very biodiverse). He and his group do not know all the answers, he admits, but are thinking and talking about many of these practical questions. Some of them relate to the timeframe one considers. Perhaps in the longer period of hundreds of years, some of the need for such forest interventions will dissipate, he suspects.

Would there be any ability to harvest in a limited and sustainable way some amount of wood from such forests, perhaps as a way for the forest to generate revenue that would allow it to take care of itself? That is another interesting question that Peter and others have indeed thought about, though he is inclined to think that that kind of use of the forest—cutting down trees—would be mostly inconsistent with the idea of the forest that owns itself.

Akkerman sees a growing number of examples of the application of the rights of nature as well as older, re-discovered precedents. In our conversation he pointed to the setting aside of the Haagse Bos, or the Hague Forest, by William of Orange in the Redemption Act of 1576, which forbade the cutting down of trees and assured nearby communities that the land would forever remain a forest, which it mostly has.

Conclusions

The challenges cities face today in protecting trees and expanding their forest canopies will require new thinking and new approaches. Luckily there are many new and emerging tools, initiatives, and ideas that cities can draw from. This chapter has explored this variety of new ways of thinking about trees, as well as the new tools and technologies that might be used to support urban forests. Many of these ideas can be combined and applied together in a city to better protect trees. I have argued for example that we need financial incentives that better reflect the importance of trees and the significant ecological and other benefits they provide us. This could happen in lots of ways but should represent a significant ramping-up of the modest forms of "treebates" that some cities currently provide. Local property tax systems should reflect and take into account the collective benefits of trees—the more trees preserved on a site the lower the property taxes should be. Cities ought to consider developing and applying a system of direct financial subsidies, especially to protect older, larger trees. Providing a property owner an annual payment to preserve or protect a large tree on site would be an unusual step in most places, but a wise investment, and a potentially powerful incentive to ensure that important trees are not thoughtlessly cut down.

There are as well many ways that cities can help trees to become a more important part of urban life. Support for new tree and forest rituals might be one way, and there is growing evidence of the value of marking life events with tree planting (birth, death, commitments to place). Tree rituals can help to make life in cities more meaningful and to seamlessly integrate our sense of past, present, and future.

Care for trees and urban forests will in turn require new and stronger emotional connections and connectedness. This might happen through the very low-tech practice of naming or individualizing trees in one's neighborhood. It might also be assisted through the use of many new and emerging digital technologies. The "internet of trees," used as a broad label for many of these, holds potential not only for more effective urban forest management (e.g., knowing when trees need to be watered) but also to be able to monitor and learn more about the physical and biological qualities of trees. As we learned from earlier chapters, creative digital strategies such as giving every tree in a city an email address or deploying a digital tree map that allows you to find and learn about the specific trees around us where we live can serve as powerful instruments for cultivating this connection and caring.

This chapter has emphasized the importance of cultivating a new ethic of trees and forests, one that especially moves beyond a narrow view of trees as essentially private property, to do with as one casually pleases, to understanding them as living creatures imbued with a high degree of public value. And as we learn more about trees and forests as complex, interconnected living systems, exhibiting even what forest ecologist Suzanne Simard refers to as form of sentience, many of us are inclined to recognize the inherent moral worth of trees and forests, and the possibility of a forest that "might own itself." It is early in the history of these ideas, and many of the others discussed here, but the time is certainly now to engage in spirited discussion about which combination of these new ideas, initiatives, and technologies will help us in advancing the vision of canopy cities.

It is also important to recognize that many of the topics discussed here are potentially reinforcing. Changing the economic incentive structure in cities, economically rewarding protection of trees, applying a sizable economic penalty when trees are cut down, serves to send an important moral signal about trees. So also does the practice of commemorating important life events by planting and celebrating trees.

Notes

1 Samantha Beattie, "Toronto Agreed to Buy a Home to Save a 250-year-old-tree. Now the Seller Wants a Higher Price," *CBC News*, July 28, 2021, found here: https://www.cbc.ca/news/canada/toronto/red-oak-tree-toronto-court-1. 6119248, accessed May 17, 2022.
2 Ryan Patrick Jones, "This 6-Year-Old Girl Is Raising Funds to Save a Majestic Toronto Oak Tree," *CBC News*, May 14, 2018, found here: https://www.cbc. ca/news/canada/toronto/this-6-year-old-girl-is-raising-funds-to-save-a-majestic-toronto-oak-tree-1.4662809, accessed May 20, 2022.

3 See Tim Beatley, "The Burrowing Owls of Marco Island," *Biophilic Cities Journal*, 2022, found here: https://static1.squarespace.com/static/5bbd32d6e 66669016a6af7e2/t/62cc4b2020414c7b1e9cbecc/1657555748165/Burrowing +Owls.pdf, accessed May 22, 2023.

4 See Mark Oswold, "Old Cottonwood Sparks Santa Fe Debate," *Albuquerque Journal*, July 7, 2015, found here: https://www.abqjournal.com/609235/sena-plaza-cottonwoods-fate-not-yet-decided.html, accessed August 17, 2022.

5 Mark Oswold, "Big Tree Being Removed From Historic Santa Fe," *Albuquerque Journal*, August 7, 2019, found here: https://www.abqjournal.com/1350823/big-tree-being-removed-from-historic-santa-fe-courtyard.html?amp=1, accessed August 17, 2022.

6 Robin Wall Kemmerer, *Braiding Sweetgrass: Indigenous Wisdom, Scientific Knowledge and the Teaching of Plants*, Penguin Books, 2013.

7 See "Robin and her Tree" found here: https://treeboston.org/tree-stories/, accessed June 5, 2023.

8 See the map of the finalists here: https://www.treeoftheyear.org/about-the-contest, accessed May 30, 2023.

9 More about the tree here: "Queen Elizabeth's Oak," found here: https://www.historic-uk.com/HistoryMagazine/DestinationsUK/Queen-Elizabeths-Oak/, accessed Oct 20, 2021.

10 Ibid.

11 Angi Gonzalez, "Meet the Queens Giant: New York City's Oldest Resident," *Spectrum News NY1*, June 17, 2019, found here: https://www.ny1.com/nyc/all-boroughs/news/2019/06/17/meet-the-queens-giant–the-oldest-tree-in-all-of-new-york-city#

12 Frank Hopper, "Native "Wall Warriors" march and gear up for a Standing Rock-style fight," *Navajo-Hopi Observer*, February 19, 2019, found at: https://www.nhonews.com/news/2019/feb/19/native-wall-warriors-march-and-gear-standing-rock-/

13 See "Runestones of Sweden," *Mashable*, found at: https://mashable.com/2017/01/07/runestones-of-sweden/#1xoPXScy6mqG

14 "Atlanta's Champion Trees," found here: https://www.treesatlanta.org/resources/atlantas-champion-trees/, accessed April 27, 2023.

15 And a chance to say goodbye to a tree being removed and to also learn about trees being planted as replacements. See "Rotterdam Bids Farewell to Dead Trees with QR Codes," *The Mayor*, January 11, 2022, found here: https://www.themayor.eu/en/a/view/rotterdam-bids-farewell-to-dead-trees-with-qr-codes-9720, accessed May 22, 2023.

16 "QR Codes Used to Encourage Citizens to Adopt Neighborhood Trees,": DC Department of Transportation, April 12, 2012, found here: https://www.springwise.com/qr-codes-hung-trees-enable-easy-adoption/, accessed May 22, 2023.

17 See E.B.D. Clabough, E. Kaplan, D. Hermeyer, T. Zimmerman, J. Chamberlin, and S. Wantman. "The Secret Life of Baby Turtles: A Novel System to Predict Hatchling Emergence, Detect Infertile Nests, and Remotely Monitor Sea Turtle Nest Events," *PLoS ONE* Vol. 17, No. 10, 2022: e0275088. https://doi.org/10.1371/journal.pone.0275088, found here: https://journals.plos.org/plosone/article?id=10.1371/journal.pone.0275088, accessed May 22, 2023.

18 Ben Green, *The Smart Enough City: Putting Technology in Its Place to Reclaim Our Urban Future*, MIT Press, 2020.

19 Katie Weeman, "Tree Sensors Track Urban Growth, Flowering and More" *Phys.org*, May 3, 2022, found here: https://phys.org/news/2022-05-tree-sensors-track-urban-growth.html, accessed May 22, 2023.

20 Mike Moore, "How Singapore Is Turning to Tech to Keep Tabs on its Trees," *Tech Radar*, November 27, 2022, found here: https://www.techradar.com/news/how-singapore-is-turning-to-tech-to-keep-tabs-on-its-trees, accessed May 22, 2023.

21 Ibid.

22 Hamid Zurqani of the University of Arkansas has been recently quoted as saying: "Drones have become an essential tool for forest management and forest monitoring … This drone will be utilized to characterize key indicators of urban health and investigate statistical relationships between forest health, such as individual forest trees and species, and vegetative indices derived from the drone imagery. It will also be utilized to calculate the trees' heights and crown widths and use this information to calculate how much carbon is held in above-ground biomass." Lon Tegels, "Drone Latest Tool for Forest Health," *Arkansas Democrat Gazette*, May 19, 2022, found here: https://www.arkansasonline.com/news/2022/may/19/drone-latest-tool-for-forest-health/, accessed May 22, 2023.

23 More about the company here: https://phys.org/news/2022-05-tree-sensors-track-urban-growth.html. To watch a video about these drones, see: https://interestingengineering.com/innovation/drones-plant-40000-trees-daily, accessed May 22, 2023.

24 See "Robot Rangers Fuse Nature With Technology," designboom.com, Sept 25, 2021, found here: https://www.designboom.com/technology/robot-rangers-segev-kaspi-designed-to-rehabilitate-forests-09-25-2021/, accessed May 22, 2023.

25 More about these traditions can be found here: "Acorn and Oak," https://www.elon.edu/u/traditions/symbols-sayings/, accessed May 22, 2023.

26 Leena Dahal, "Elon's Oak-Giving Custom Lives On," *Elon News Network*, May 18, 2014, found here: https://www.elonnewsnetwork.com/article/2014/05/elons-oak-giving-custom-lives, accessed May 22, 2023.

27 Patrick Wright, "International Students Plant Oak Saplings, Leaving Physical Legacy," *Today at Elon*, May 29, 2020, found here: https://www.elon.edu/u/news/2020/05/29/international-students-plant-oak-saplings-leaving-physical-legacy-at-elon/, accessed May 22, 2023.

28 Sami Grover, "Why I Gift My Daughters a 'Forever Forest' Every Birthday and Holiday Season," Treehugger, December 12, 2022, found here: https://www.treehugger.com/forever-forest-journal-tree-gift-6890706, accessed May 22, 2023.

29 See One Tree Planted, "2021 Planting Report," found here: https://issuu.com/onetreeplanted/docs/one_tree_planted_2021_annual_report?fr=sZWUxYzQ1ODYyMTY, accessed May 22, 2023.

30 "Green Your Holidays with a Future City Tree," San Francisco Environment Department, found here: https://sfenvironment.org/adopt-a-living-tree-for-christmas, accessed May 22, 2023.

31 San Jose Christmas Tree Rentals, https://plantman.com/live-christmas-tree.html, accessed May 22, 2023.

32 Sherryn Groch, "What Do These Sacred Trees Tell Us About Aboriginal Heritage in Australia?" *The Sydney Morning Herald*, October 31, 2020, found here: https://www.smh.com.au/national/what-do-these-sacred-trees-tell-us-about-aboriginal-heritage-in-australia-20201030-p56a0g.html, accessed May 22, 2023.

33 Sarah Hollister, "Placenta Burial Traditions," found here: https://placentarisks. org/wp-content/uploads/2018/09/Placenta-burial-rituals-from-around-the-world-handout.pdf, accessed May 22, 2023.

34 Nina McConigley, "Rooting a New Life under a Juniper Tree," *High Country News*, December 1, 2022, found here: https://www.hcn.org/issues/54.12/essays-rooting-a-new-life-under-a-juniper-tree, accessed May 22, 2023.

35 Interview with Lee Webster, March, 2019.

36 https://carolinamemorialsanctuary.org/, accessed May 22, 2023.

37 See an excellent short film here: https://carolinamemorialsanctuary.org/, accessed May 22, 2023.

38 E.g. see Angela Hutti, "Controlled Burn at North St. Louis Prairie Fuels Good Growth, Eliminates the Bad," November 9, 2022, found here: https://fox2now.com/news/missouri/controlled-burn-at-north-st-louis-prairie-fuels-good-growth-eliminates-the-bad/, accessed May 22, 2023.

39 Constant Mecheut, "Wild and Wilde: At Celebrity Cemetery, Nature Takes a Starring Role," *New York Times*, December 28, 2022, found here: https://www.nytimes.com/2022/12/28/world/europe/paris-pere-lachaise-cemetery-.html, accessed May 22, 2023.

40 Interview with Lee Webster, 2019.

41 Emma Newcombe, "From Cemeteries to Suburbs: How a Romantic Movement Reshaped America," *Governing*, December 22, 2022, found here: https://www.governing.com/community/from-cemeteries-to-suburbs-how-a-romantic-movement-reshaped-america, accessed May 22, 2023.

42 See Arbor Day Foundation, found here: https://shop.arborday.org/commemorative?gclid=CjwKCAiAzKqdBhAnEiwAePEjkmv2XZN_rjh9l8So9SSTSwW3Wgzc6LwSM-Lx2qSAJ6Ir1CIrA9H2NhoCqscQAvD_BwE, accessed May 22, 2023.

43 See https://www.betterplaceforests.com/, accessed May 22, 2023.

44 Thomas Farragher, "A Berkshire Forest Blossoms Into a New Kind of Graveyard," *Boston Globe*, January, 25, 2022, found here: https://www.bostonglobe.com/2022/01/25/metro/berkshire-forest-blossoms-into-new-kind-graveyard/, accessed May 22, 2023.

45 E.g. Caitlin Doughty, "If you want to give something back to nature, give your body," *New York Times*, December 5, 2022, found here: https://www.nytimes.com/interactive/2022/12/05/opinion/human-composting-new-york.html, accessed May 22, 2023.

46 Meghan Shinn, "9-11 Memorial Trees," *Horticulture*, September 12, 2011, found here: https://www.hortmag.com/gardens/9-11-memorial-trees, accessed May 18, 2023.

47 E.g. "Tree Dedication to Memorialize Victims of King Soopers Shooting," March 31 2021, found at: https://www.colorado.edu/fm/2022/03/17/tree-dedication-memorialize-victims-king-soopers-shooting, accessed May 18, 2023.

48 Trevor Nace, "Icelandic Forest Service Recommends Hugging Trees Since You Can't Hug People," *Forbes*, April 14, 2020, found here: https://www.forbes.com/sites/trevornace/2020/04/14/icelandic-forest-service-recommends-hugging-trees-since-you-cant-hug-people/, accessed May 18, 2023.

49 Richard K. Broughton et al., "Long-Term Woodland Restoration on Lowland Farmland Through Passive Rewilding," *PLOS One*, June 16, 2021, found here: https://journals.plos.org/plosone/article?id=10.1371/journal.pone.0252466, accessed May 22, 2023.

50 Presentation of James Anderson and Chris Gillespie, to the Biophilic Cities Network, November 16, 2022.

51 See Patrick Greenfield, "Revealed: More than 90% of Rainforest Carbon Offsets by Biggest Certifier Are Worthless, Analysis Shows," *The Guardian*, January 18, 2023, found here: https://www.theguardian.com/environment/ 2023/jan/18/revealed-forest-carbon-offsets-biggest-provider-worthless-verra- aoe, accessed May 18, 2023.

52 For a full explanation of this case see Lisa Song and James Temple, "A Nonprofit Promised to Preserve Wildlife. Then it Made Millions Claiming it Could Cut Down Trees," *ProPublica*, May 10, 2021, found here: https://www. propublica.org/article/a-nonprofit-promised-to-preserve-wildlife-then-it-made- millions-claiming-it-could-cut-down-trees, accessed May 18, 2023.

53 Chris Davis, "Austin Buys Carbon Credits from New Tree Planting Program as Carbon Neutrality Goal Looms," *KXAN*, found here: https://www.kxan.com/ news/local/austin/austin-buys-carbon-credits-from-new-tree-planting- program-as-carbon-neutrality-goal-looms/, accessed May 18, 2023.

54 "The Rainforest" found here: www.brooklynbridgeforest.com

55 "The Partnership" found here: www.brooklynbridgeforest.com

56 "Preserve a Landmark. Protect a Rainforest. Cultivate a Global Partnership," at www.brooklynbridgeforest.org

57 See https://www.goodmeat.co/about

58 "How 'Lab Grown' Meat Could Help the Planet," *CNN*, July 15, 2022.

59 Dryden Quigley, "Azalea Springs Housing Plans Raise Concerns After Narrowly Approved by the City Council," *WHSV3*, January 12, 2023., found here: https:// www.nbc29.com/2023/01/12/azalea-springs-housing-plans-raise-concerns- after-being-narrowly-approved-by-city-council/

60 "Public Trust Doctrine," found here: https://www.watereducation.org/ aquapedia/public-trust-doctrine, accessed May 24, 2023.

61 See "TNC, Funding Trees For Health: Finance and Policy to Enable Tree Planting for Public Health," September 23, 2017, found here: https://www. nature.org/en-us/what-we-do/our-insights/perspectives/funding-trees-for- health/, accessed July 11, 2022.

62 Aldo Leopold, *A Sand County Almanac*, Ballantine Books, 1949, p.263

63 Mary Oliver, "The Black Walnut Tree," in *New and Selected Poems, Volume One*, Beacon Press, 1992, p.201.

64 Stephen Kellert, *Birthright: People and Nature in the Modern World*, Yale University Press, p.xiv.

65 Email communication with Jake Van Yahres, September 13, 2023.

66 See "The Tree That Owns Itself," found here: https://www.visitathensga.com/ things-to-do/attractions/the-tree-that-owns-itself/, accessed May 22, 2023.

67 Junior Ladies Garden Club of Athens, "Our History," found here: https:// juniorladiesgc.org/our-history/, accessed May 22, 2023.

68 Christopher Stone, *Should Trees Have Standing? Toward Legal Rights for Natural Objects*, Los Altos, CA: William Kaufman, 1974.

69 Alex Traub, "Christopher Stone, Who Proposed Legal Rights for Trees, Dies at 83," *New York Times*, May 28, 2021, found here: https://www.nytimes.com/ 2021/05/28/us/christopher-stone-dead.html, accessed May 22, 2023.

70 Stone, 1974, p.11.

71 See Katie Surma, "Ecuador's High Court Affirms Constitutional Protections for the Rights of Nature in a Landmark Decision," *Inside Climate News*, December 3, 2021, found here: https://insideclimatenews.org/news/03122021/ ecuador-rights-of-nature/, accessed June 1, 2023.

72 Erik Stokstad, "This Lagoon Is Effectively a Person, says Spanish Law that's Attempting to Save it," *Science*, September 29, 2022, found here: https://www.science.org/content/article/lagoon-effectively-person-says-spanish-law-s-attempting-save-it, accessed May 22, 2023.
73 C. R. Stuart-Ukin, "Quebec's Magpie River Becomes First in Canada to be Granted Legal Personhood," *National Observer*, February 24, 2021, found here: https://www.nationalobserver.com/2021/02/24/news/quebecs-magpie-river-first-in-canada-granted-legal-personhood, accessed May 16, 2023.
74 Interview with Jessica den Outer, January 23, 2023.
75 Interview with Peter Akkerman, January 30, 2023.
76 The Foundation website can be found here: https://www.bosdatvanzichzelfis.nl/en/

10

LIFE IN A CANOPY CITY

Conclusions and Future Directions

There are few things we can do in the design and management of cities that will have greater importance and impact than saving trees and growing out a city's forest canopy. As this book has tried to make abundantly clear, the benefits of trees and forests to urban residents are immense and multifaceted. Urban challenges, including increasing heat and reduced outdoor time, are overlapping and posing significant health risks. Older, larger trees offer significant physical and mental health benefits, but fewer residents and reduced social contact further undermine health and wellbeing.

Trees in cities are a remarkable superpower: at once addressing multiple urban challenges. Planting trees, protecting existing older trees, and expanding urban forests will provide catalytic and multiplying benefits—they will cool urban environments and help to clean the air, thus making cities healthier, but they will also encourage residents to spend more time outside, and in turn help to make spaces and neighborhoods more sociable, in turn enhancing health.

Trees and forests help to ensure that cities are habitats for many other forms of life. This is a good and right thing to do, and we are duty-bound, I believe, to share urban spaces. But the results will also deliver and pleasure, and mystery and joy, to urban life. I have often said that the abundance and quality of native birdsong is an important measure of a good city. And I believe it strongly. But this will be next to impossible in cities with few trees and limited forest canopy.

It is hard to overstate the value of trees and forest from the perspective of awe and wonder, things often in short supply in heavily developed cities.

DOI: 10.4324/9781003377344-10

Dacher Keltner makes a strong and convincing case for the benefits of awe. To him, awe can derive from many different human experiences, including moral beauty, art and design, and music, among others. But nature is an especially potent source, and trees and forests are remarkable activators of awe. I think again about the reactions I had when first I glimpsed a giant sequoia. But I had a similar feeling of vastness and awe when I experienced massive and old cottonwoods on the streets of Santa Fe. Or the remarkable live oaks dripping with Spanish moss in New Orleans. Or closer to home, and on an almost hourly basis, the large beautiful White Oak tree just steps from my office Virginia. We want and need these trees and forests around us, to bring us pleasure certainly, but to invoke awe and to encourage us to step beyond our narrow self-interests, and to feel connected to others and a larger world.

Trees may be one of the most essential ingredients in bringing about a more just and caring world. The scholarship of Keltner and others shows that in the presence of trees and forests, we are able to tame what he calls "mean egotism."[1] We are likely to be more empathetic, more interested in the future, and more willing to tolerate others and work together to create a better future. As Keltner says so eloquently: "By quieting the nagging, self-critical, overbearing, status-conscious voice of our self, or ego, and empowering us to collaborate, to open our minds to wonders, and to see the deep patterns of life."[2]

Trees, moreover, have the power at once to connect us to the past and to the future. They are in this way time-travelers of the first order, and they inject into urban life a deep time dimension, connecting the distant past with the far-away future. Especially now, when we need the ability to think longer term, the protection of existing trees, and the planting of new forests, represents both a symbolic commitment to the future and a tangible step in bringing about a more flourishing future for all.

Trees Are an Answer to Multiple Problems Facing Cities

I have tried to make the hopefully convincing case in this book that trees and urban forests are an especially potent response to the most pressing challenges facing cities today. Trees and urban forests are an unusually effective strategy for addressing the growing problem of urban heat. There are many promising ideas cities can use in addressing urban heat, including reflective paints and pavement, and design to stimulate and capture cool breezes, but none of these ideas are likely to be as effective as trees will be.

In addition to climate change, cities face a variety of other pressing social challenges. These include raising suicide rates and mental health problems associated with social isolation, among others. We are drawn to trees, and they soften urban environments in so many important ways. We are more likely to want to be outside where there are trees and nature. Trees and forests have a civilizing effect. Trees are a universal and positive life-force in cities around the world and while the climate and specific species will vary, the benefits are remarkably similar (see Figure 10.1).

FIGURE 10.1 Trees in the Public Plaza in Oaxaca, Mexico. Photo credit: Tim Beatley.

American society, especially it seems, entails an emphasis on indoor living, and indoor working. It is frequently said that we spend more than 90% of our day inside and this may be further evidence of the need to bring trees and nature inside, which we are increasingly doing. But we also need to work on getting more of us outside, and spending more of our day outdoors. It is where we will experience trees and urban forests to the fullest extent. Ironically, efforts to preserve older, existing trees and planting new trees will be a key part of the strategy for propelling us outdoors. We know in urban neighborhoods and environments that have more trees and nature we are likely to spend time outside. But we also need to spend more time outside to fully appreciate the importance of trees in the first place. Here I see value in cultivating, especially in the US, more of an outdoor culture.

In Nordic countries there is a word—*friluftsliv*—or "open-air life" that describes this culture. The word comes from an 1859 Ibsen poem, "On the Heights," and can be defined as "embracing nature and enjoying the outdoors as a way of life."[3] As Swedish writer Linda Akeson McGurk notes, the word is meant to include experiences of nature that are "non-competitive and non-motorized."[4] Not so much a set of practices or activities in fact, McGurk calls it a "broader philosophical lifestyle." "Slow nature," is another descriptor she uses.

It is interesting to imagine what it will take to truly bring about an outdoor forest culture for those who essentially and perhaps exclusively occupy cities. It is close to the idea put forth by Stephen Kellert that while we carry the innate impulse to affiliate with trees and nature, that is biophilia, this really requires education and development.

How does this happen and can we really imagine the emergence of an urban forest culture? I hope so. McGurk, in her descriptions of her own childhood, suggests the importance of growing up around family and others who are mentoring, gently guiding, and nudging in favor of an outdoor life:

When I was growing up in Sweden, friluftsliv was more or less consciously passed down to me from the adults in my life—grandparents, parents, early childhood educators, teachers, and other caregivers. When I was little, they made sure I got to play outside every day, rain or shine. They taught me the names of our local plants and wildlife and passed down old stories and legends that made the forests come alive with magic. They gave me the time and space to wade in shallow creeks looking for frog spawn and fill my pockets with rocks and other special treasures. They let me nurture my sense of wonder while observing the frenzied activity of an anthill and trying to grasp the concept of infinity while gazing up at the Milky Way. Thanks to them and the culture I grew up in, the outdoors was the constant backdrop against which my childhood played out. McGurk, 2022, p. xvi.

Cities can support the development of an outdoor culture in many different ways. Planning and organizing abundant public events in outdoor spaces is one thing. Winter cities like Edmonton, Canada, have made special strides to encourage outdoor life during those months when residents might rather curl up on the couch and watch TV or surf the web. There is a winter strategy there, and an annual ice castle that becomes a must to see and visit, for example. Public forest walks, outdoor concerts, creating many accessible public spaces that make it possible and easy to reach the water, woods, and other wild spaces around a city all can play a role.

The global pandemic has opened up new spaces and opportunities for rethinking where and how we actually work. Many workers have resisted the call to return to the conventional office setting and perhaps we are at an unusual moment when working outside or out of doors will become a legitimate option for many. Trees and forests in cities will need to play a big part, and we will need to think creatively about how and where workspaces could be integrated and designed in. Could this new interest in working outside in turn become a major force on behalf of urban trees and forests—planting more trees, protecting existing older trees, and working toward expanding urban canopies? I think so.

I return to some of the creative ideas offered by engineer-artist Natalie Jeremajenko and specifically her ideas and prototyping of tree workspaces. One notable example is the TREExOFFICE designed in collaboration with the architecture firm Tate Harmer. Able to accommodate up to eight workers, this circular outdoor structure was built around a tree in London's Hoxton Square. There are no limits to what kinds of creative outdoor work structures might be imagined and built, perhaps in collaboration and with funding from companies whose workers will be able to use them.

Perhaps it will be something simpler even than this. Just having a grove of trees to sit under becomes an opportunity. I recall more than a decade ago spending time as a visiting professor on the campus of the University of New South Wales, where there were several very popular courtyards where large trees provided beautiful, shaded spaces for students, staff, and faculty to be. There were some benches, but mostly people sat directly on the ground, alone or in circles. There was a popular coffee vendor nearby and food to be had not far away as well. It was not an issue in this location, but maybe we hand out blankets or cushions for those who want to have something to sit on. And of course Wi-Fi access will be important, and maybe other amenities that might enhance the experience and the attractiveness of spending at least a part of the work day outside among the trees.

Efforts like the now international Outside Office Day are helpful as well. Started in 2019 by the Dutch-based organization Nature Desks, it is now in its fifth year. Usually the date is mid-June (the 15th of June in 2023) and the variety of organized outdoor spaces and programming is impressive. I am taken by some of the simple solutions—the table and chair taken into the nearby forest, for instance. There is no question that we are in the midst of a global shift in the ways we work and that this presents opportunities to spend more of our day outside—including visiting, sitting under, listeningy, and working around trees and forests. Urban forests and trees as places of work will necessarily grow in importance, I think (and hope). A 2022 *Gallup poll* found that the vast majority of respondents (a sample of more than 8,000 workers) did not want to return to the office full time, and half or more workers already split their time between office and home (and other spaces).[5] The evidence suggests that worker productivity (and creativity) go way up when working outside, so there are clear incentives for companies to encourage these new and emerging patterns of outdoor work. This new mode of working, stronger especially after the global pandemic, offers an incredible opportunity to activate new forms of civic life and to invigorate public spaces of various kinds, and also an opportunity to reconnect to the trees and nature around us.

Every Planner a Forester?

Every urban planner should also work to become an urban forester. Urban planners can no longer afford to restrict their domain to the design of buildings, roadways, hard surfaces, and conventional infrastructure. They will increasingly need to think about nature and work to protect and restore that nature within cities, and this will especially include trees and forests.

Trees are going to have become an even more essential element in the design of civic spaces in cities, and this will be important for several reasons. In a world of increasing heat, few will want to spend time in the public realm without the shade and cooling provided by trees. But it is more than that; it is the combining of the power of seeing, listening to, and experiencing nature, in combination with the awe experienced by what Durkheim called "collective effervescence," highlighted in Keltner's book, *Awe*.

The pedestrianized Downtown Mall in my home city of Charlottesville is a living example of how this can work. Here, families and residents of all ages co-occupy these forested pedestrian spaces, sitting, eating, playing, but most especially strolling (see Figure 10.2). Collective strolling is made possible (tolerable and enjoyable) by the tree clusters and sheltering canopy, and during the very busiest times it becomes an experience close to the collective effervescence described previously.

FIGURE 10.2   The Charlottesville Downtown Mall (with more than 60 mature trees). Image credit: Tim Beatley.

What this in turn means is that the ways we teach urban planners today must change. Rarely is a course about trees and forests to be found at all in a planning curriculum, nevertheless something that all students would be required take.

There are some academic programs where a minor in urban forestry is available, and I am hopeful that the next generation of planners will see such credentials as necessary and valuable. It is common today that planning students, partly with the goal of increasing their employability, seek out additional credentials—for instance LEED certification through the US Green Building Council. It is useful for planning students to be thinking about green and sustainable buildings, but perhaps even more important for them to think about trees and forests.

Professional urban planners will increasingly need to make connections and build bridges with other professions and disciplines involved in pro- tection and management of trees and forests. These parallel tree-friendly professions include civil engineering, landscape architectures, parks and recreation, and public health among others. Trees and forests must be a key element of any local health plan, given their importance in supporting physical and mental health. There is a need for a more organized discussion and coordination of the tree protection and advocacy efforts of these various professions.

New Visions of Cities

This book has argued that to fully protect trees and forests, and to achieve a vision of future cities where canopies are healthy and expanding, and where older, larger trees are adequately protected, will require a different way of thinking. And especially a different way of thinking about those trees. It is not simply a technical solution or even giving trees more weight in the political system, though that would certainly help.

The new urban tree ethic we need is multi-pronged. It includes the need to expand our understanding of trees beyond simply seeing them as another form of private property. Rather, we must understand trees and forests as profoundly imbued with a civic character. We must see them as providing essential public services—shade and coolth, air quality, the absence of noise, and birds and birdsong, among many other things. And when there is a proposal to cut them down we must see such an act as diminishing and harming the larger public good.

It is also part and parcel of a new and different view of cities, one that embraces the need for nature at the center. Trees and urban forests represent the best hope for an urban future in which people get to experience more than the occasional visit to a park but rather get to live immersed in nature. This immersive nature also helps to cool the city and to allow it to be more resilient.

Part of this vision for future cities is (again) about more outdoor, and nature-connected, lives and lifestyles. The global pandemic has resulted in the loss of many lives, but if there is a silver lining it is a rediscovery of the value and importance of nature. We want and need trees and forests in our lives—not far away, only to be visited on a holiday—but close by, indeed all around us. The vision of forest urbanism recognizes, moreover, that trees and forests and urban development need not, and must not, be seen as antithetical. Future urbanism can include trees and forests, as we have seen in this book, and it can happen in many ways: new housing projects can integrate existing older trees, buildings can make room for trees on balconies and rooftops, shifting our mobility patterns away from cars will create opportunities to convert surface parking and roadway space to trees and forests. We will still plant trees along streets, but we can look for places to insert wilder, more *transgressive* forms of forest—forests with multiple layers and with ecological complexity and biological diversity; forests that are messy and extend (grow) into the many spaces of a city. There are many exciting ways in which making room for trees and forests will help to create new and more interesting kinds of cities in the future!

New Partnerships and Alliances for Trees and Forests

There is a need for new alliances and to develop a more robust set of community organizations that support and defend a city's trees and forests. Luckily there is now a robust and growing set of community-based organizations in most cities focused on trees and forests.

Examples include Friends of the Urban Forest, in San Francisco, Tree Action Seattle, Trees Atlanta, and Trees Pittsburgh. In many cities there are multiple organizations focused on tree planting and tree protection. Some focus on advocating for and actively working to protect existing trees, for instance *The Last 6000 Campaign* in Seattle. Casey Trees in Washington and the Philadelphia Horticultural Society (PHS) in that city are also wonderful examples of non-governmental organizations working to extend and expand urban canopies. Others have focused more on tree planting, and organizations like Urban Releaf, in Oakland, California, on the important work of planting trees in low-canopy, underserved neighborhoods.

One clear message is that cities must do more to protect the older trees that already exist. They are doing the lion's share of the ecological work, and allowing the cutting of an old tree in exchange for planting new seedlings makes little sense and is a dangerous strategy.

It is ironic that as I am writing this there is a frenzied battle to save the 200 or so Sequoia trees threatened by wildfire in Yosemite National Park. It is clear why we would want to do everything humanly possible to save these trees—they are unusually old and unique and ancient living members of our community of life. Yet we seem unable to attach a similar reverence to the very old trees that can be found in almost any city. A reverence and deep appreciation for these ancient trees would do much to shift the careless and ambivalent ways we treat them today.

There is a tremendous opportunity to enhance the quality of urban life by recommitting to trees and forests. And along with trees will come birds and many other forms of life that deliver pleasure, delight, and meaning.

While the benefits of trees and forests are great indeed, cities face a variety of challenges in the future, many discussed in the chapters of this book. Some of these challenges are financial: how to find the fiscal resources to pay to expand urban forests and to undertake the care and management they will need. Others are environmental and climatic: climate change especially will be further complicating the stresses urban forests face and the management challenges facing cities. Many of the challenges, however, have more to do with societal values and the current ways we think of and treat trees and forests. In cities the prevailing attitude is still, mostly, that they are expendable and replaceable, with a status only slightly higher than street furniture.

We need to explore various partnerships in cities to protect more trees and to expand the canopy. Many organizations and groups will have a parallel mission that could and should complement trees and forests.

Real estate developers and builders are a significant and important constituency group if we are to save trees in cities and work to grow urban canopies. In the American planning and development system, developers are largely the ones in the driver's seat. And while they respond to market demand and market pressures, real estate products tend to be conservative: building homes and developments that have been popular and successful in the past. The bank and financial institutions in turn reinforce these conservative tendencies.

Developers need convincing that trees—preserving existing trees, and planting new ones—will add positive value to their products and will add to the sales price and profit margins of their projects. It is surprising that even today developers must be convinced of the value of trees. I have provided several examples of very successful real estate projects that have made trees and forests central design elements. These have included the examples of Oak Forest Terrace, in North Charleston, South Carolina, and Bryant Heights in Seattle.

Further efforts to educate and enlist the help of the development world will be needed, in combination with stronger tree protection codes as well. The irony for me is mass timber designs have become very popular, very fast and now prominent examples can be found in cities around the world. This is a positive trend certainly, but unfortunately, we need equally to see real estate projects and developments that emphasize protection and creative integration of existing living trees as well.

We badly need more positive examples of creative development and real estate projects that demonstrate that trees can be protected while at the same time increasing density and housing. This was one thing I heard in my discussions with the leaders of *The Last 6000*. They could identify but one project in Seattle that could serve as this kind of positive example.

A big part of every city's challenge will be is to take tangible steps to ensure the older trees around one's home stay protected and are allowed to age. The trees around the house I grew up in were an essential part of my life from the very beginning. I didn't give their long-term status much thought, of course, certainly not when I left home to go off to college and to start my adult life. The house and lot were later sold and a new owner (I did not realize until later) had other ideas, replacing the home with a massive new structure and taking out many of the trees I grew up with. It did not even occur to me that I needed to take steps to ensure the survival of those trees. But perhaps conditions could have been placed on the sale of the property that would have protected some or all of the trees. This kind of thing seems rare or unlikely but could be a regular part of estate planning and the drafting of wills. I think homeowners, and their extended families, ought to give (and be expected to give) more thought to what will happen to the trees.

As more communities around the country begin to densify and adjust their zoning codes to allow more of the so-called "missing middle" housing, there is both opportunity and peril. There is peril in the sense that careless accommodation of density will lead to the loss of trees, no doubt. But there is an opportunity to more creatively design and fit-in this new density in ways that acknowledge and seek to protect especially the older trees that exist. Partly this means there must be a process for careful review to ensure that every consideration is given to protecting trees.

In 2022, a design class at the University of Virginia, led by Professor Jonah Coe-Scharff, explored some of the design possibilities for fitting in new missing middle housing created by changes in Charlottesville's comprehensive plan and zoning ordinance. This led to an exhibition and report identifying five specific ideas for "inclusive infill housing." The models built to illustrate these options were striking for the prominence they gave to trees. This work showed, among other things, the absolute importance of visualizing how single-family neighborhoods can grow and densify but also protect the forest canopy that will be so crucial to ensure these neighborhoods are cool and habitable.

I think it is time to challenge homeowners to think more expansively about their own duties to protect the trees around them in their neighborhoods. There are examples of course of homeowners who have voluntarily nominated trees on their property for heritage status and protection (where such options exist) but these seem few and far between. Homeowners should be challenged, however, to think more carefully about this and to take steps to ensure the trees they care about today will continue to exist and be cared for after they move away or are otherwise not in the home.

I believe we need new models for how tree service companies and tree cutters operate in cities. In line with new efforts to better control rogue tree cutting in cities like Seattle, perhaps what we need to explore and promote is an entirely new kind of approach that might expand the mission of such companies to better emphasize tree conservation and tree planting, and in the process maybe take away some of the inborn incentives that lead to too quick and too aggressive cutting down of trees.

A tree company operating in the Atlanta suburb of Alpharetta may provide an example of a different approach. Recently this private company issued a press release announcing the launch of an urban reforestation program.[6] How refreshing that a private company in the business of trimming and cutting trees also gives attention to growing and even expanding the city's canopy. We ought to hold such companies to a higher standard—for every client or homeowner who hires them to cut something down, there should be the expectation that an equal amount of time is spent explaining how and why the homeowner should be replacing and/or expanding the number of trees planted on their property.

New Tree and Forest Roles

Among the other partnerships most promising are those that might involve the religious communities, especially in the U.S. A vivid personal example for me of the connections (or lack thereof) between religion and urban forests came in the early months of the global pandemic, when the world began to more fully appreciate the essential benefits of nature. On one of my walks home I discovered a very large and quite old tree being cut down in the yard of a prominent church. After several days this ancient tree was completely gone. Its age and seniority did not protect it unfortunately, and at the time I thought the actions of the church to be ungodly and reflecting the antithesis of the stewardship obligations some churches have professed.

Churches I could be strong champions for urban tree and forest conservation. I am reminded of E.O. Wilson's efforts to reach out to mainstream religions, like Christianity, in an effort to establish common ground for supporting efforts to protect and conserve biodiversity. The religious and secular alike, Wilson believed, ought to agree on the importance of protecting "creation" and trees and forests certainly fall under that umbrella.[7]

Artists will also need to play a role in embracing trees and expanding local canopies. From the digital tree installation inside the Cleveland Clinic to the many efforts to sketch, paint and visually convey the deeper essence and meaning of trees, artists are essential in efforts to raise the importance of trees in cities. We must also enlist artists and the art world in reimaging cities as forests. In Leeuwarden, in the Netherlands, the Walking Trees have helped to spark remarkable conversations about the importance of nature and have helped to visually demonstrate how trees and forests can reshape and reimagine cities.

In every city there will be anchor institutions that can be enlisted to help with protecting and expanding the forest canopy. From churches to schools to museums and libraries there are natural (though perhaps not always obvious) connections between trees and the mission of these community anchors. I recently received an email alert from the Cleveland Museum of Natural History, for example, imploring me to: "BECOME A MEMBER, PLANT A TREE!" It turns out that as a part of their membership drive they will plant one native red oak sapling for every new membership or renewal. These plantings will take place in the Mento Marsh State Nature Preserve, on the edge of Lake Erie, and a State of Ohio designated natural area.[8] It is a habitat the museum had a major role in protecting and restoring, going back to the 1960s. Natural history museum's are an example of an important kind of community anchor institution, and many cities are lucky to have them. And they can become significant partners and catalysts for trees and urban forests.

We must also work to enlist the medical and healthcare world on behalf of trees and forests. Medical professionals and the healthcare industry are natural partners and allies. There has been a growing awareness about the health-enhancing role of nature, and this is a very positive development. Physicians, nurses, and healthcare professionals are increasingly enlisting nature as a response to physical and especially mental health challenges. The connections to health is a compelling argument for protecting and ex-panding forests in and around cities. The medical world, broadly defined, represents then an especially important set of allies for tree conservation and should be the focus on future efforts at partnerships and collaborations. This has taken many forms. In some cases it is about prescribing nature experiences—a visit to a park, time spent outside, and of course a walk in the forest.

Several years ago, I had the opportunity to present at an annual con-ference organized by Moda Health Care, one of the largest healthcare companies operating in the Pacific Northwest and Alaska. I learned then that the company had been providing funding for the Nature Conservancy to plant trees in Oregon. That made tremendous sense. Though Moda might not directly reap the benefits of healthier policyholders, it made complete sense that healthcare providers and health insurance companies would want to utilize and leverage the remarkable health and wellness benefits of trees and forests.

Recent research on the mental health benefits, especially of trees and forests, leaves little doubt about their value as a profound form of health-enhancing action. Again, there are few interventions in cities that will deliver better health, that will serve up more potent forms of medicine, than trees and forests. More trees will reduce mortality and extend life expectancies as several studies have demonstrated.

Tree planting (and especially protection of the trees and canopy that already exist) will, moreover, be incredibly cost-effective and a very wise investment for cities. When the economic value of extended lives is calculated and fully taken into account the benefits-to-costs ratio are high.[9]

Teachers and educators of all sorts are another constituency we need to better reach as we prioritize trees and forests. There are so many impressive ways that a visit to a forest can help to shape a young life (and eventually a career path). Trees and forests should be seen as ways to make schools and school yards cooler and more hospitable places to learn and play (and especially in underserved neighborhoods) but they must also be seen as essential equipment for learning. One example of this is the Chattahoochee Hills Charter School, just outside of Atlanta, Georgia. Designed as a series of separate structures to facilitate physical activity, the kids come prepared with their boots and coats to spend much of the day outside. There is a forest adjacent to the school that becomes a place in which to take one's math or science homework.[10] A school forest becomes not only an outdoor classroom but also an environmental science living laboratory.

One of the more pressing societal challenges has to do with our aging population. Trees and forests are of course not the only, or even the primary, response in addressing the needs of those (including myself) who are in or are entering elderhood. I believe the gray wave represents an unusual opportunity to advance the vision of tree cities. As we get older, we also have the tendency to confront our mortality and to think more about what kind of legacy we might leave to our families and communities. Erik Erickson's famous stages of development suggests that as we grow older, we are likely to see the value and importance of giving back, of leaving something for the future—what Erickson calls *Generativity*.[11] A more natureful city, and more trees and forests, would seem a natural expression of this kind of legacy-thinking. And the more legacy-minded we are the more we might be encouraged as we get older to take specific steps to protect the ancient trees around us that we love (perhaps bequeathing that old white oak, and ensuring it isn't cut down by the next owner of our home after we pass along). As the number of legacy-minded elders grows I feel this could be a powerful and important force on behalf of urban tree conservation.

As many thousands retire and look for meaningful experiences in the later years of their lives, connections with nature will become more important. Planting and caring for trees may be an especially gratifying activity for older residents and one that responds especially well to instincts at this age (I am feeling them) of wanting to give back, of wanting to do things today that will leave a positive legacy. Few things that one can do now provide as much assurance of delivering benefits for the future as protecting trees and planting new ones.

Organizations New and Old: Top Down or Bottom-Up?

One of the key conclusions of this book is that no single individual, agency or other entity can be fully responsible for protecting and advocating for a city's forest. No one office or organization will be able to protect and grow a city's forest canopy. It will require the concerted work of many different organizations and entities.

Garden clubs have a role, as has been evident in the stories from Pittsburgh to Athens, Georgia, where the Junior Garden Club took such an important role in protecting the tree that owns itself, but more broadly became an advocate for tree planting and forest conservation in that city. So also do organizations like the Rotary Clubs, with civic-minded chapters all over the US and perhaps beyond.

We need to support and cultivate in cities a healthy and robust network of community organizations working on trees and canopy issues. Even in cities with strong centralized urban forestry programs and strong tree codes, these grassroots groups will be necessary partners in pushing for implementation, enforcement and for lobbying for adequate annual budgets. And we will need groups who are able to spring into action when trees are threatened. The importance of both citywide tree protection codes and standards but also the need to empower and facilitate more grassroots and neighborhood-based conservation.

There are many positive examples to cite in our Biophilic Cities, including TreePeople in Los Angeles, TreeFolks in Austin, and Friends of the Urban Forest in San Francisco. (see Figure 10.3). They vary in size and focus but they together comprise a remarkable and growing network of non-governmental organizations that are working on behalf of trees and forests.

Encouragingly, many of these organizations are specifically motivated by historic inequities and discrimination in cities that also manifest in terms of inequitable distribution of trees and canopy. These include groups like Southside Releaf, in Richmond, Virginia, that have worked to overcome the long-standing impacts of historic redlining in that city, through tree planting in underserved areas. Residents of the southside of that city can experience significantly lower life expectancies (up to twenty years shorter) so the implications of these historic patterns of discrimination and spatial segregation are significant indeed.[12] These equity-centered groups have been important partners and advocates for changing the vision and policy in cities, for instance in the case of Richmond the adoption of a new comprehensive plan (Richmond 300) that establishes minimum canopy targets for all neighborhoods.

FIGURE 10.3   Friends of the Urban Forest Has Been an Important Advocate for Trees in San Francisco. Photo credit: Tim Beatley.

The volunteer group in Seattle, *The Last 6000*, mentioned in earlier chapters, is one example of how a small core group of citizens can mobilize on behalf of trees.[13] I spoke via Zoom with four of the leaders of this group not only about the challenges they face but also the promise of citizen-based tree advocacy and protection. This small group formed in response to a citywide tree canopy study that estimated there were 6,000 old-growth trees left in the city (more specifically defined as trees 30 inches or more DBH). Each of these core members also comes to the group with personal experiences and stories of older trees lost along the way. One co-founder, Barbara Bernard, described a parcel across the street with a grove of trees that was clear cut in a redevelopment project. "We all have stories of trees we have loved and lost," says Sandy Shettler, who describes them as "ghosts" that are remembered and remain present. All these leaders share the surprise with just how easy it is to clear away even very large trees. The group is active politically and advocates for stronger city tree protections, but has also initiated a process for discovering and collecting more specific information about the remaining 6,000 trees. Citizens are asked to submit information about them and so far they have received information about 1,400 majestic trees, as Jim Davis another *Campaign* co-founder, refers to them. So far the collection process has been old-school, asking citizens to fill out a hard-copy form, providing information about the tree and its location. While they plan to digitize the forms, Jim says they are not sure they want to publish a map for fear it will provoke more loss of trees (he refers tongue in check to the trees being in a witness protection program).

Despite the small number of core leaders this group has been a mighty force. There are 1,700 followers on their Instagram account, and they are able to activate and mobilize tree enthusiasts in other parts of the city. Jim Davis estimates that at least 500 people have been active or involved in some way in their efforts, attending an educational tree walk or submitting a tree record. They have been dogged critics and watchdogs, pointing out where the city's existing tree policies are not being enforced or implemented. The on-the-ground community activism of *The Last 6000 Campaign* shows the power of work by neighbors to actively work to care for specific threatened trees that they know and love. As Shettler says it is important to think about the broader policy picture, of course, but that it is important to realize the way people connect with and care about specific trees.

Identifying and training more local and neighborhood "tree activists" is definitely a goal Jim Davis tells me, and definitely a need. More eyes on the trees, and more dedicated residents who have vested interests in seeing their beloved trees protected. Davis says Seattle could use another 500, and be composed ideally of more diverse faces.

Forward-looking cities could help in the cultivation of these tree activists and neighborhood tree watchdogs. Many cities offer courses and training aimed at educating the public and creating a pipeline of people who can serve on city committees and planning commissions and ideally some who will run for public office. Trees and urban forests could be a standalone training but at the least must be a prominent and important unit in any such courses. It is a small investment of resources with potentially immense payback. Cultivating the next generation of informed, capable and dedicated tree citizens is an essential step that every city, whatever its size, should undertake.

And like the resilience of a diverse urban forest, a city will benefit from many different kinds of citizen groups and initiatives. While *The Last 6000 Campaign* has a focus on activating citizens to protect threatened trees, other groups could focus on planting new trees and work to restore trees and forests and this could also happen at the neighborhood scale. One example is the small but dedicated group working to grow and restore a forest "patch" in the Langdon neighborhood of Washington, DC. Begun as an effort to tag and protect the young sapling trees emerging in the Langdon Park, and to ensure that park maintenance crews did not mow them down, it has gradually evolved into a larger citizen science project, aimed at studying how forest patches grow and develop under different planting and care scenarios.

One of the leaders of this effort, Allison Clausen, notes the importance of actual residents of the surrounding neighborhood working on the patch. "We love that we get to meet our neighbors while we're doing this work. We love that we feel ownership. It's like our space. And that's really motivating. We get more permission to do it because we're neighbors, we are citizens directly connected to this park."[14]

Part of this is a matter of practicality—it's just easier to donate one's time to a project that is nearby, and those involved will likely care more about these nearby spaces. And as Clausen suggests, it is likely that residents of the neighborhood will have more credibility and more success in taking on the restoration of forests that are perceived to be part of their neighborhood.

The Langdon Patch experiment has also generated useful information about reforesting small spaces in a city. They learned for example that the many vines and invasives that often take over sites can be effectively managed through a combination of turning the soil and using mulch. And there may be a highly effective model for growing the urban forest here. Clausen notes that in just two years the patch has produced (and they have tagged) more than 700 young trees. Whether they will all or mostly survive to maturity is unclear but compared to the citywide target of planting 10,500 trees, this is a pretty impressive outcome.

The Langdon Forest Patch has now even been officially acknowledged by the DC government and even a sign now erected on site, suggesting the possibility that similar neighborhood-based forest restoration and research projects could happen elsewhere in the city. It is another wonderful way to cultivate love of (and knowledge about) the trees around us. And a way to grow a neighborhood constituency who has invested considerable time and energy in planting and studying these trees and will likely do whatever is necessary to make sure the patch persists. The idea of forest patches further conveys the sense that trees and groves can popup in many places in cities, on often unlikely small parcels, and that a city of many patches could present the opportunity to connect them together to, as a method of stitching back together the forest that was there before the city existed.

And in the case of the Langdon Forest Patch, it has become a "place," a destination for neighborhood residents, something added to their mental maps, and now a forested trail where one did not exist before. It has added a forest amenity to the neighborhood already enjoyed by many residents and that will further reinforce connections with and care for nature and trees. This is viewed by many as a model that could apply throughout the city.

And Clausen at the end of her presentation offers suggestions about how citizens can get involved. "Find a wood near you," she advises. "There are these little bits of woods all over the city and they could use your help."

It is commonly observed that people will not care about or come to the aid of people or things that they don't know personally or don't have a personal connection with. In this regard, cities need to further support and invest in the many possible ways that residents can experience, enjoy, visit, explore and otherwise get to know the trees and forests around them. This could happen in many different ways, as we have seen, from regular tree walking tours, to online tree maps, to facilitated neighborhood discussions about the trees around them that are most beloved. And some of these ideas represent the chance to generate jobs and income, for instance by giving guided tree and nature walks.

New Forms of Canopy Politics

Elected officials need to be held accountable for their decisions about trees and forests and the modest level of funding they typically receive. New York City Mayor Eric Adams who had earlier committed to spending at least 1% of the city's budget on parks and nature, including trees, seems recently to have backed away from this commitment. In the face of rising crime, failing bridge and road infrastructure, and a growing opioid drug crisis, it may not be surprising that urban trees and forests receive less attention and are a lower priority for funding. That is unfortunate of course (and short-sighted in the sense that many of a city's underlying problems and crises are at least in part addressed by trees and forests).

Can we push for city council candidates to campaign on a platform that protects and expands nature, and especially trees and forests? What would a successful model of local tree politics look like? One obvious element would be the placement of trees and forests on the local political agenda; that seems a minimum step (see Figure 10.4). Those campaigning for elected office ought to explicitly mention trees and forests in campaign platforms. In New York City there is a growing consensus that the city should adopt the goal of planting another one million trees, and these kinds of specific platform commitments by candidates I think can be very helpful.

Taking specific steps to strengthen the law around trees in a city is another important category of candidate commitment. Protection for existing older trees and ambitious canopy targets ought to make their way into campaign literature and speeches, though this rarely seems to be the case in most city politics today. A new form of tree politics will in turn require local advocacy organizations to prioritize trees and make these tree priorities clear to local candidates.

FIGURE 10.4 Tree Protection Ought to Be a Prominent Element of Any City Council Candidate's Campaign Platform (here an example of a campaign sign from Charlottesville, VA). Photo credit: Tim Beatley.

Every city needs, I believe, a strong tree code. Just as we seek clear rules about what is acceptable when it comes to driving and road safety or building codes and construction standards, to name just a couple of areas where individual behavior is legitimately constrained for a larger public interest, so also should we be clear about rules that pertain to trees and forests. These rules must be clear about when and under what circumstances it is permissible to cut down trees. And there must be a clear and defensible process for making decisions, a process that seeks to reduce conflicts of interest and works to strengthen transparency. Cities must do better. The recent example of Sheffield, in the UK, is instructive as a cautionary tale of a city that sought to delegate decisions about tree management to a private company, kept the public in the dark, and whose elected officials failed to assume accountability when the public rejected the excessive tree cutting that occurred there.[15]

Much of the political support for trees, and the adoption and enforcement of stronger tree codes, will in turn be activated or undergirded by an engaged and passionate tree-loving citizenry and through the activism of a robust set of community groups and tree advocacy organizations. Fierce citizen opposition to tree cutting in the case of Sheffield helped to stop it there. In the case of New York City, it has been organizations like New Yorkers for Parks that have been pushing Mayor Adams to follow through on his 1% pledge.[16]

The politics of trees and tree conservation are important to think about more carefully than we have. What are the most effective arguments that will sway residents about the need to protect existing older trees, and to plant new trees to expand the canopy?

For many, trees are viewed as inanimate objects, passive elements of the urban scene. That is changing, as I have noted in earlier chapters, thanks to the scientific research about the agency and social life of plants and trees. And there is a lens of wonder and awe that we need to continue to develop and apply in our conservation efforts, as exemplified by the activism and writing of people like Suzanne Simard and Diana Beresford-Kroeger. The magic and majesty especially of older trees will motivate many of us (I hope) (Figure 10.5)

Perhaps for many the fact that trees are so essential to the survival of other forms of life that we love (such as birds) will be more convincing, or a more compelling form or argument or motivation for conservation. I hope so. My colleague Peter Newman from Curtin University in Western Australia, and others, have made this argument in a recent paper in the journal *Urban Science*: "The charismatic appeal of iconic species, such as the Black Cockatoo, have the potential to function as catalysts for education, and for winning widespread community support, to implement such measures as [nature-positive design and regeneration]."[17]

"The risk of extinction for the Black [Cockatoo] highlights the crisis being faced in the Perth region much more saliently, for example, than the statistics of tree loss might." That may often be true and true for many urban residents. Saving the trees may become more important when they are associated with saving the birds we love.

If this is true, it may be another example of what botanists have called "plant blindness," accurately observing that animals have tended to garner more attention and support than the more ubiquitous and less animated plants that support them. Here I think of the important work of University of Delaware entomologist Douglas Tallamy (mentioned earlier in the book), who has demonstrated compellingly that if you care about birds, you must also care about trees and native trees especially; the ones that provide the caterpillars that sustain most nestlings. He told me it was not by accident that the cover of his book, *Nature's Best*

FIGURE 10.5 A Large Cottonwood Near Albuquerque, New Mexico. Photo Credit: Anneke Bastiaan.

Hope, sports the image of a robin holding a caterpillar in its beak: perhaps he is right that we will tend to be more motivated by love of our favorite birds than love of trees (though I believe many of us do indeed love trees as well)[18] (Figure 10.5).

New Ways to Fund Trees and Forests

Cites today face a host of interlocking challenges as they work to protect trees and forests as well as expand their canopies. Many of these have already been highlighted in previous chapters. Authority and responsibility for urban trees and forests is often fragmented and, as the new Boston Forest Plan describes it, subject to a "patchwork "of agencies and organizations. Frequently cities lack a central coordinating office or position, though increasingly cities are appointing a city forester or similar position to serve this function.

As this book has made clear (hopefully), there are few economic investments with as clear and quick a return as those focused on expanding tree canopies. The benefits are many and long-lived and the costs by comparison relatively modest. Yet, paying for programs to plant trees and perhaps especially their long-term care will require the dedication of significant local resources.

Cities will need to explore a variety of creative methods for funding, to be sure. Some cities have created tree funds that allow developers and others to pay in-lieu fee payments when trees are cut down. Such funds can serve as a helpful beginning, but tapping into other sources will be necessary. Utilizing methods that seek to capture some amount of the increase in property values due directly to the planting of trees (such as a tool like tax increment financing) can be part of the answer. Because trees and forests are such an important way to enhance health and wellbeing and to respond to climate change there is a good argument to be made that a dedicated sales tax or other dedicated revenue stream is needed and appropriate.

Carbon offsets will continue to represent one funding stream for cities, though a healthy degree of skepticism and more rigorous methods for ensuring that the carbon sequestration benefits are real and additive are in order. Local efforts at tree planting and canopy protection will certainly result in reduced carbon emissions, and some credit, where demonstrable, for these effects should also be given. It has been suggested, for instance, that carbon offsets might be a legitimate source of funds, in cases like Pittsburgh's City Steps, where investments in pedestrian systems and network of treeways can be shown to reduce car usage.

Partner cities in our Biophilic Cities Network are utilizing a variety of other creative funding tools and ideas. Washington, DC, for instance, has pioneered the use of Environmental Impact Bonds,[19] while San Francisco has been using Green Benefit Districts,[20] and at a more regional level the use of a $12 per year parcel tax that will generate $25 million over twenty years to support wetlands restoration efforts.[21] Radically re-growing a city and metro area's canopy is an ambitious project and worthy of an ambitious funding scheme. It is worth noting that in the case of the parcel tax, created through a referendum (Measure AA), more than 70% of voters supported the measure. I would expect similar support for trees and forests, perhaps even greater support, if the public is engaged in a discussion and educated about the benefits that will accrue.

Crowd-sourcing platforms might also be a source of tree funding especially for neighborhood level projects. Platforms like IOBY (In Our Backyard), that emphasize local investors who also have the chance to participate in the undertaking of the project (e.g., the planting and watering of trees), are especially promising.[22] But such sources may not take the place of larger sources of public funding such as a dedicated sales tax.

In cities in the Global South funding for trees is even more tenuous though equally needed. The example of Freetown the Treetown (Freetown, the capital of Sierra Leone), is instructive. With seed funding from the World Bank, the city has embarked on a program of citizen-initiated tree planting and tree care, where income is generated for participants. The World Bank describes the campaign this way: "Community-based growers use its TreeTracker app on locally available smartphones to create a unique geotagged record, or ID, for each new tree planted—including a photo. Growers revisit each seedling periodically, to water and maintain, verify and document the plant's survival—and receive mobile money micropayments for their efforts."[23] The program helps on several levels to address tree loss at the city's periphery, as a result of population growth pressures, and to help advance the city's goal of planting a million new trees.

Technology: A Help or a Hindrance in Canopy Cities?

The Freetown story is perhaps one example of how modern digital technology can assist cities in expanding their canopies. Apps that utilize geocoded information can serve as an effective mechanism for funding and monitoring the health of trees planted in cities perhaps faraway (to offset carbon emissions for example). More generally, the internet of trees offers the potential to more effectively manage urban trees and forests. Online tree maps, like New York's, have the potential to raise awareness of the health of neighborhood trees and to organize and coordinate watering and other care.

Some of the ways that our cell phones make urban life more convenient can also help to make more room for trees. The example from an earlier chapter of parking apps is a case in point. If such apps make it easier to locate parking spaces and more efficiently utilize urban parking it may be possible to reduce redundancy, there may be opportunities to convert some of these urban spaces to trees and forests. And our digital technology may help us to utilize many alternatives to automobiles (for instance by finding an electric scooter) and in that way may also assist in transitioning spaces in cities to forests.

But ubiquitous cell phones and constant surfing of the internet also represents a major distraction from the natural world, and there remains a challenge especially for how children relate to nature and trees. Cities ought to be open to utilizing these technologies to educate and connect but there is also the need to disconnect, to create experiences and opportunities for kids (and adults) to enjoy trees and forests in ways that allow them to be fully attentive.

These smart city ideas and technologies can be used to better grow and protect trees and forests. For example, moisture sensors on trees can become important ways to at once concern diminishing water and ensure that trees survive periods of drought, that in an era of climate change will undoubtedly become more common and last for longer periods. This is part of what Nadina Galle has called the "Internet of Nature," and she espouses a decidedly optimistic view of the power and utility of these kinds of digital technology. Efficiently monitoring and caring for trees will be a clear benefit.

Digital technologies might also be helpful in making trees and forests a more central part of urban life. Our smartphones can help guide us to the nearest forest and can deliver immense amounts of information about the trees we are experiencing (and the birds and other life we find and hear in those trees). They can also become tools in helping remind us when we need to take a break and go outside for an awe walk.

How to take advantage of our digital technology but also manage its downsides will remain a challenge in canopy cities. Much of this digital technology is distracting and distancing, of course, and therein lies the delicate balance cities face. Simply putting down our phones, turning off our laptops, will often be the best thing we can do to appreciate the trees and nature around us, as well as the friends and family around us sharing these experiences. But there is no putting the genie back in the bottle, and so it probably behooves us to continue to find ways to harness the power of these technologies on behalf of canopy cities.

Other Obstacles to Forest Urbanism?

There are many real obstacles to protecting and expanding the forest canopy in cities, many of which have already been discussed—limited staffing and financial resources, and relatively lower political importance are key ones.

Some of the other obstacles we face have to do with overcoming what some think of as the negatives of living in and around trees. In hurricanes and storm events we see lots of imagery of downed trees, trees that have fallen on roofs and that have blocked roads. It is natural that news organizations and film crews want to capture the most dramatic images following a storm. In my own leafy neighborhood, there is a tendency to blame trees whenever the lights flicker, and the power goes out. We never see the downed trees that are so commonly blamed. It seems that trees often are the go-to scapegoat when it comes to such disruptions.

But the truth is quite different, and we need to collectively tell a different story. Trees, rather than being a liability, actually help to shield and protect our homes and communities. The occasional and relatively infrequent downed tree is a small price to pay for the beauty, cooling, presence of birds and wildlife, and many other benefits we gain. The CEO of the https://www.arborday.org/ Arbor Day Foundation, Dan Lambe, was recently quoted in a Sacramento news story offering a helpful, contrary view about trees and weather:

"[Trees] help to slow the strong winds, buffer communities, buffer homes, buffer neighborhoods from the strong wind events, " says Lame. "They also intercept heavy rain waters to reduce and slow runoff and flooding."[24]

Trees and their root systems are often seen as a disruptor of sidewalks and at the municipal level are things that need to be trimmed, watered, and otherwise cared for. Many questions often arise about how to keep trees alive in drought, something we will experience more frequently everywhere. Ironically trees and forest canopy are a big part of the answer to water shortages. In Los Angeles, for example, some 40% of the city's water needs are provided from groundwater, levels of which are in decline despite the immense amount of rainwater that falls there. That city's stormwater infrastructure (including famously the LA River) has emphasized the collection and fast movement of floodwater from land to sea. More trees and forests and a more permeable urban landscape that would capture, slow down and allow this water to percolate back into the ground is increasingly understood as an essential part of what is needed.[25]

The need to design and plan for resilience, moreover, increasingly suggests the value and importance of trees and forests. The inclusion of a ten-acre forest in one of Baltimore's "resiliency hubs," suggests an important role for trees in helping cities and their residents become more resilient in the face of growing stressors including heat and severe weather events.

The underlying social and cultural contexts must also be acknowledged. In the US especially a culture of excessive individualism makes it increasingly difficult to institute any form of collective regulation or restraint. I have argued here that every city should adopt and faithfully enforce a strong tree code that protects existing trees and also mandates new tree planting. Yet, in a country and culture where it has become difficult even to curtail personal ownership of military-grade machine-guns, telling a homeowner they may not be able to cut down an ancient tree seems increasingly unlikely. A political culture that eschews and denigrates the concept of enforcing common rules and standards aimed at protecting the larger public interest will necessarily work against canopy-cities.

But the most significant obstacle of all may be the current ways we tend to perceive trees and forests. They are, especially in cities, often seen at best as desirable amenities but expendable in the face of a more pressing need or project. They are often understood to have a status not much higher than street furniture. That is changing in many places, of course, as I have discussed throughout this book, but perhaps the change is not occurring fast enough. To fully advance the vision of canopy cities advocated here will require a wholesale renovation of our ethical posture to trees and forests—a recognition of their intrinsically collective and civic dimensions, for one. That we all benefit from trees and that deciding to cut down a tree is not simply an individual decision, but a choice with larger collective implications. In short, we need a new ethic that understands the public nature of trees, but also recognizes their intrinsic moral value. Movement in the direction of recognizing the rights of nature is encouraging and suggests that this expanded ethic may indeed be possible.

In the shorter term we will need, in cities, to muster the multitude of arguments in support of trees and forests. We will need to continue to emphasize their many ecological and social benefits, their value in enhancing resilience and wellbeing, and how they add immeasurably to our quality of life and to the beauty and civility of our cities.

Notes

1 Dacher Keltner, *Awe: The New Science of Everyday Wonder and How It Can Transform Your Life*, Penguin Press, 2023.
2 Keltner, *Awe*, p.xx
3 Linda Akeson McGurk, *The Open-Air Life: Discover the Nordic Art of Friluftsliv and Embrace Nature Every Day,* TarcherPerigree Books, 2022, p.xiv.
4 Ibid, p.xv.
5 See Gallop, "Returning to the Office: The Current, Preferred and Future State of Remote Work," *Gallup*, August 31, 2022, found here: https://www.gallup.com/workplace/397751/returning-office-current-preferred-future-state-remote-work.aspx, accessed June 2, 2023.
6 "Tree Service Alpharetta Launches Urban Reforestation Program To Increase Canopy," *Digital Journal*, March 31, 2023, found here: https://www.digitaljournal.com/pr/news/tree-service-alpharetta-launches-urban-reforestation-program-to-increase-tree-canopy-cover-by-30-, accessed May 15, 2023.

7 See E.O. Wilson's wonderful book *The Creation: An Appeal to Save Life on Earth*, W.W. Norton, 2007.

8 More about Mento Marsh here: https://www.cmnh.org/mentor-marsh, accessed May 22, 2023.

9 See Geoffrey H. Donovan et al., "The Association Between Tree Planting and Mortality: A Natural Experiment and Cost-Benefit Analysis," *Environment International*, Vol. 170, December 2022, found here: https://www.sciencedirect.com/science/article/pii/S0160412022005360, accessed May 10, 2023. According to their findings: "Using US EPA estimates of a value of a statistical life, we estimated that planting a tree in each of Portland's 140 Census tracts would generate $14.2 million in annual benefits (95% % CI: $8.0 million to $20.4 million). In contrast, the annual cost of maintaining 140 trees would be $2,716–$13,720."

10 For a short documentary film about this school see: https://www.biophiliccities.org/chattahoochee-hills-charter-school-film, accessed May 16, 2024.

11 E.g. see Leonardo Becchetti and Davide Bellucci, "Generativity, Aging and Subjective Well-being," *International Review of Economics*, Vol. 68, 2021, 141–184, found here: https://link.springer.com/article/10.1007/s12232-020-00358-6, accessed June 2, 2023.

12 See Southside Releaf, here: https://www.southsidereleaf.org/, accessed May 22, 2023.

13 Interview with Barbara Bernard, Jim Davis, Jessica Dixon, and Sandy Shettler, The Last 6000, August 8, 2022.

14 Presentation by Allison Clausen, Sponsored by Capital Nature, November 17, 2022.

15 Helen Pidd, "Sheffield City Council Behaved Dishonestly in Street Tree Row, Inquiry Finds," *The Guardian*, March 6, 2023, found here: https://www.theguardian.com/uk-news/2023/mar/06/sheffield-city-council-behaved-dishonestly-in-street-trees-row-inquiry-finds, accessed May 18, 2023.

16 See James Barron, "One Percent of the Budget for Parks? A Bargain Says a Nonprofit," *New York Times*, March 13, 2023, found here: https://www.nytimes.com/2023/03/13/nyregion/one-percent-of-the-budget-for-parks-a-bargain-says-a-nonprofit.html#:~:text=Budget%20for%20Parks%3F,A%20Bargain%2 C%20Says%20a%20Nonprofit,says%20New%20Yorkers%20for%20Parks, accessed April 14, 2023.

17 Giles Thomson et al., "Nature-Positive Design and Development: A Case Study on Regenerating Black Cockatoo Habitat in Urban Developments in Perth, Australia," *Urban Science*, 2022, found here: https://pdfs.semanticscholar.org/e6a7/aa412d4056e6b9c48a62c5300a8d010a7469.pdf, accessed May 22, 2023.

18 Douglas Tallamy, *Nature's Best Hope: A New Approach to Conservation That Starts in Your Yard*, Timber Press, 2020.

19 "DC Water: Environmental Impact Bond," found here: https://www.quantifiedventures.com/dc-water, accessed May 16, 2023.

20 "Green Benefits Districts," found here: https://www.sfpublicworks.org/GBD, accessed May 15, 2023.

21 "San Francisco Restoration Authority Measure AA," found here: https://myparceltax.com/sfbay/#:~:text=This%20special%20tax%20is%20levied,tax%20may%20not%20be%20prepaid, accessed May 15, 2023.

22 "About IOBY," found here: https://Ioby.org/about, accessed May 16, 2023.

23 "FreetownTheTreeTown campaign: Using digital tools to encourage tree cultivation in cities," found here: https://blogs.worldbank.org/sustainablecities/freetownthetreetown-campaign-using-digital-tools-encourage-tree-cultivation, accessed May 16, 2023.

24 Dan Lame quoted in "'It's emotional for communities': Trees fall in Sacramento, continue protecting city from severe weather," January 14, 2023, found here: https://www.abc10.com/article/tech/science/environment/trees-fall-continue-to-protect/103–71012427-0d03–4aa7-ab9e-97073536f7f7

25 E.g. see Paul Thornton, "Imagining How a Wilder L.A. and its Rivers Would Have handled this Rain," *LA Times*, January 14, 2023, found here: https://www-latimes-com.cdn.ampproject.org/c/s/www.latimes.com/opinion/story/2023-01-14/rain-in-los-angeles-before-concrete-rivers?_amp=true

BIBLIOGRAPHY

Arlington County, VA, 2020. *Pentagon City Sector Plan*, Arlington County, VA.

Astell-Burt, Thomas, Michael A. Navakatikyan, and Xiaoqi Feng, July 23, 2023. "Why Might Urban Tree Canopy Reduce Dementia Risk?" *Health & Place*, Vol. 82.

Beatley, Timothy, 2020. *The Bird-Friendly City: Creating Safe Urban Environments*, Washington, DC: Island Press.

Beatley, Timothy, 2011. *Biophilic Cities: Integrating Nature in Urban Design and Planning*, Washington, DC: Island Press.

Beatley, Timothy, 2014. *Blue Urbanism: Connecting Cities and Oceans*, Washington, DC: Island Press.

Beatley, Timothy, 2017. *Handbook of Biophilic City Planning and Design*, Washington, DC: Island Press.

Beatley, Timothy and J. D. Brown, 2021. "The Half-Earth City," *Environmental Law and Policy Review*, William & Mary Law School, Vol. 45, Spring 2021, 775–819.

Beresford-Kroeger, Diana, 2011. *The Global Forest: Forty Ways Trees Can Save Us*, Penguin Books.

Beresford-Kroeger, Diana, 2019. *To Speak for the Trees: My Life's Journey from Ancient Celtic Wisdom to a Healing Vision of the Forest*, Random House Canada.

Bliss, Laura, 2022. *The Quarantine Atlas: Mapping Global Life Under COVID-19*, New York: Black Dog and Leventhal Publishers.

Browning, W. D., C. O. Ryan, and J. O. Clancy, 2014. *14 Patterns of Biophilic Design*, New York: Terrapin Bright Green llc.

Browning, William and Catherine Ryan, 2020. *Nature Inside: A Biophilic Design Guide*, RIBA Publishing.

Cabanek, Agata, Maria Elena Zingoni de Baro and Peter Newman, 2020. "Biophilic Streets: A Design Framework for Creating Multiple Urban Benefits," *Sustainable Earth*, July.

Chi, Dengkai et al., May 11, 2022. "Residential Exposure to Urban Trees and Medication Sales for Mood Disorders and Cardiovascular Disease in Brussels, Belgium: An Ecological Study," *Environmental Health Perspectives*, Vol. 120, No. 4, found here: https://ehp.niehs.nih.gov/doi/10.1289/EHP9924, accessed June 4, 2022.

Cox, Daniel T. et al., 2017. "Doses of Neighborhood Nature: The Benefits for Mental Health of Living with Nature," *BioScience*, Vol. 67, No. 2, February, 147–155.

Donovan, G. H. and D.T. Butry, 2010. "Trees in the City: Street Trees in Portland, Oregon," *Landscape and Urban Planning*, Vol. 94, 77–83.

Donovan, Geoffrey H. et al., December 2022. "The Association Between Tree Planting and Mortality: A Natural Experiment and Cost-benefit Analysis," *Environment International*, Vol. 170.

Ferrini, Francesco, Cecil C. Konijnendijk van den Bosch, and Alessio Fini, 2017. *Routledge Handbook of Urban Forestry*, Routledge Press.

Gagliano, Monica, 2018. *Thus Spoke the Plant: A Remarkable Journey of Groundbreaking Scientific Discoveries and Personal Encounters with Plants*, North Atlantic Books.

Gatti, Robert Cazzola, et al., 2022. "The Number of Trees on Earth," *PNAS*, Vol. 19, No. 6.

Guo Fengyi, et al., 2023. "Autumn stopover hotspots and multiscale habitat associations of migratory landbirds in the eastern United States," *PNAS*, Vol. 120, No. 3.

Haviland-Jones, Jeannette, Holly Hale Rosario, Patricia Wilson, and Terry R. McGuire. January-December 2005. "An Environmental Approach to Positive Emotion: Flowers," *Evolutionary Psychology*, Vol. 3, No. 1. https://doi.org/10.1177/147470490500300109COPYCITATION

Immergluck, Dan, 2022. *Red Hot City: Housing, Race and Exclusion in Twenty-First Century Atlanta*, University of California Press.

Kaplan, S. 1995. "The Restorative Benefits of Nature: Toward an Integrative Framework," *Journal of Environmental Psychology*, Vol. 15, 169–182. 10.1016/0272-4944(95)90001-2

Kellert, Stephen, 2014. *Birthright: People and Nature in the Modern World*, New Haven, CT: Yale University Press.

Kellert, Stephen and Elizabeth Calabrese, 2015. *The Practice of Biophilic Design*, John Wiley Press.

Kellert, Stephen R., Judith Heerwagen, and Martin Mador, 2008. *Biophilic Design: The Theory, Science and Practice of Bringing Buildings to Life*, John Wiley Press.

Keltner, Dacher, 2023. *Awe: The New Science of Everyday Wonder and How It Can Transform Your Life*, Penguin Press.

Kemmerer, Robin Wall, 2013. *Braiding Sweetgrass: Indigenous Wisdom, Scientific Knowledge and the Teaching of Plants*, Penguin Books.

Konijnendijk, Cecil C. 2019. *The Forest and the City: The Cultural Landscape of Urban Woodland*, Springer.

Leonardi, Cesare, 2019. *The Architecture of Trees*, Princeton Architectural Press.

Lewis, Hannah, 2022. *Mini-Forest Revolution: Using the Miyawaki Method to Rapidly Rewild the World*, Chelsea Green.

Locke, Dexter et al., 2021. "Residential Housing Segregation and Urban Tree Canopy in 37 US Cities," *npj Urban Sustain* Vol. 1, No. 15. 10.1038/s42949-021-00022-0

Loomis, Brandon, 2022. "'Life and Death Infrastructure': Volunteers Plant Trees for a New Phoenix 'Cool Corridor'," *AZ Central*, April 17, found here: https://www.azcentral.com/story/news/local/arizona-environment/2022/04/17/beating-heat-phoenix-plants-first-cool-corridor/7308133001/, accessed July 22, 2022.

Louv, Richard, 2008. *Last Child in the Woods: Saving Our Children from Nature-Deficit Disorder*, Algonquin Books.

Iungman , T, Cirach, M, Marando, F, Pereira Barboza, E, Khomenko, S, Masselot, P, Quijal-Zamorano, M, Mueller, N, Gasparrini, A, Urquiza, J, Heris, M, Thondoo, M, & Nieuwenhuijsen, M (2023). Cooling cities through urban green infrastructure: a health impact assessment of European cities. Lancet, 2023 Feb 18, Vol. 401, No. 10376, 577–589. 10.1016/S0140-6736(22)02585-5. Epub 2023 Jan 31. PMID: 36736334.

Luther, Erin, 2020. "Between *Bios* and *Philia*: Inside the Politics of Life-loving Cities," *Urban Geography*, 10.1080/02723638.2020.1854530

McGurk, Linda Akeson, 2022. *The Open-Air Life: Discover the Nordic Art of Friluftsliv and Embrace Nature Every Day*, TarcherPerigree Books.

Pederson Zari, Maibritt, 2017. "What makes a city 'biophilic'? Observations and experiences from the Wellington Nature Map project." In M. Aurel (ed.), *Back to the Future: The Next 50 Years*, 51st International Conference of the Architectural Science Association, pp. 1–10. Wellington, NZ.

Powers, Richard, 2018. *Overstory: A Novel*, WW Norton.

Rian, Iasef Md. and Mario Sassone, 2014, "Tree-Inspired Dendriforms and Fractal-Like Branching Structures in Architecture: A Brief Historical Overview," *Frontiers of Architectural Research*, Vol, 3, 298–323.

Simard, Suzanne, 2021. *In Search of the Mother Tree: Discovering the Wisdom of the Forest*, Knopt.

Stone, Christopher, 1974. *Should Trees Have Standing? Toward Legal Rights for Natural Objects*, William Kaufman.

Tallamy, Douglas, 2019. *Nature's Best Hope: A New Approach to Conservation That Starts in Your Yard*, Timber Press.

Tallamy, Douglas. 2021. *The Nature of Oaks: The Rich Ecology of Our Most Essential Native Trees*, Timber Press.

Texas Trees Foundation, 2017. *Urban Heat Island Management Study*, 2017, found here: https://www.texastrees.org/wp-content/uploads/2019/06/Urban-Heat-Island-Study-August-2017.pdf

The Nature Conservancy (TNC), 2017. *See TNC, Funding Trees For Health::Finance and Policy to Enable Tree Planting for Public Health*, September 23, 2017, found here: https://www.nature.org/en-us/what-we-do/our-insights/perspectives/funding-trees-for-health/

Ulrich, Roger, 1984. "View Through a Window May Influence Recovery from Surgery," *Science*, May, Vol. 224, No. 4647, 420–421.

Vander Groot, Jana, 2018. *Architecture and the Forest Aesthetic: A New Look at Design and Resilient Urbanism*, Routledge.

Williams, Florence, 2018. *The Nature Fix: Why Nature Makes Us Happier, Healthier, and More Creative*, W.W. Norton.

Wilson, E.O., 1984. *Biophilia: The Human Bond With Other Species*, Harvard University Press.

Wilson, E.O., 2017. *Half-Earth: Our Planet's Fight for Life*, Liveright.

Wolleben, Peter, 2016. *The Hidden Life of Trees: What They Feel, How They Communicate—Discoveries from A Secret World*, Greystone Books.

Wolleben, Peter, 2021. *The Heartbeat of Trees: Embracing Our Ancient Bond with Forests and Nature*, Greystone Books.

INDEX

Page numbers in *italics* indicate figures; page numbers in **bold** indicate tables.